Introduction to Environmental Management

Introduction to Environmental Management

Vierah Hulley

Delve
PUBLISHING
www.delvepublishing.com

Introduction to Environmental Management

Vierah Hulley

Delve Publishing
224 Shoreacres Road
Burlington, ON L7L 2H2
Canada
www.delvepublishing.com

Email: orders@arclereducation.com

ISBN: 978-1-77407-751-1 (Hardcover)

ABOUT THE AUTHOR

Vierah Hulley obtained her Ph.D. degree in Geohydrology from the University of the Free State, South Africa, in 2013. Her expertise is in Environmental and Earth Sciences and she has an extensive vocational background in the areas of natural resources management, environmental geochemistry, contaminated site remediation, sustainable development, and spatial analytics. She is the founder and Managing Director of HL Nexus: The Data Imaginarium; a company dedicated to optimizing natural resources management with data science.

TABLE OF CONTENTS

LIST OF FIGURES

LIST OF TABLES

LIST OF ABBREVIATIONS

ACE	Advisory Council for the Environment
CCO	chief compliance
CEE	Central as well as Eastern Europe
DNA	deoxyribonucleic acid
DPs	designated projects
EA	environmental assessment
EASO	Environmental Assessments of Sites and Organization
ECIs	environmental condition indicators
EIA	environmental impact assessment
EIAO	environmental impact assessment ordinance
EIS	environmental impact statement
EMAS	eco-management and audit scheme
EMF	environmental management framework
EMP	environmental management plan
EMS	environmental management systems
EP	environmental permit
EPIs	environmental performance indicators
GATS	general agreement on trade in services
GATT	general agreement on tariffs and trade
GHG	greenhouse gases
GMOs	genetically modified organisms
IPCC	Intergovernmental Panel on Climate Change
ISA	National Standardizing Associations
ISO	International Standard Organization
IUCN	International Union for the Conservation of Nature
LCA	life-cycle assessment
MCA	maximum credible accident
MDGs	millennium development goals

MPIs	management performance indicators
NBSAPs	National Biodiversity Strategies and Action Plans
NEMA	National Environment Management Authority
NGOs	nongovernmental organizations
OCHA	Office for the Coordination of Humanitarian Affairs
OPIs	operational performance indicators
OSCE	Office of Child Support Enforcement
OSHA	Occupational Safety and Health Administration
PCBs	polychlorinated biphenyls
PDCA	plan-do-check-act cycle
SEA	strategic environmental assessment
SEDAC	socio-economic data and applications center
SLM	sustainable land management
SNR	School of Natural Resources
TRIPS	Trade-Related Aspects of Intellectual Property Rights
UNCCD	United Nations Convention to Combat Desertification
UNCED	United Nations Conference Environment and Development
UNEP	United Nations Environmental Program
UNESCO	United Nations Educational, Scientific, and Cultural Organization
US EPA	The United States Environmental Protection Agency

PREFACE

Environmental management is the process of organizing various environmental programs to address different ecological matters affecting the world. Basically, environmental management attempts to avoid ecological disaster, helping in environmental crisis preservation, and seeking to find suitable solutions.

Environmental managers consist of a wide group of stakeholders including scholars, NGOs, policy-makers, and civil servants among others who make choices about the usage of natural resources. The resources on earth are limited; therefore, if we fail to use them appropriately they can become exhausted pretty soon. For the appropriate and practical use of resources, environmental management is essential. It's our basic role to form a precise coordination and balance between our requirements as humans and environmental procedures.

In terms of sustainable development, this process is also needed for development but without destroying or overusing natural resources, and limiting pollution and destruction of nature. Seeing the welfare of forthcoming generations, appropriate decisions concerning the safe use of the environment will become necessary. Nowadays, development has gotten to a point whereby the environment and ecology both are in crisis, in case the same is to continue, it may create a negative or disastrous outcome on the ecosystem.

Environmental management furthermore limits the threat of disasters such as flooding, earthquakes, desertification, forest fire, global warming, and transport accidents. There's a direct link between the environmental structure and these kinds of disasters, including the interactions between natural and man-made disasters. It's always important to draw a mark of the limit between modernization and the environment. For instance, if developmental needs are leading to global warming or the reduction of the ozone layer, we don't have to continue using the harmful materials leading to this effect, but must rather modify development to be more environments friendly.

This volume covers various topics of interest in environmental management. Chapter 1 gives a brief introduction to environmental management. Chapter 2 discusses the environmental issues. Chapter 3 gives an introduction to materials management. Chapter 4 focuses on the environmental impact assessment (EIA). The policies and legal angle of environmental management are discussed in

Chapter 5. Chapter 6 discusses the environmental management systems (EMS) and the ISO 14000. Chapter 7 focusess on the environmental ethics and Chapter 8 is about the World Trade Organization and environmental issues. Chapter 9 discuss the education and awareness on environmental management. Finally, Chapter 10 discusses the gender and environmental management.

Ultimately, environmental management involves everyone to some extent since all human actions eventually have environmental impacts. Nevertheless, some entities are more openly involved with resource utilization, besides some special interest associations are especially concerned with the exploitation of resources and matters concerning pollution.

As such, environmental management involves multiple stakeholders and covers a multi-disciplinary perspective. It further involves different spatial scales, varying from the regional to the global. Additionally, it involves multiple, diverse goals, such as the need to manage the direction and rate of development, to regulate resource use, minimizing environmental degradation as a result.

Additionally, environmental management might be practiced by people or groups holding divergent- and even openly opposing-views, like it may be the situation when environmental managers working for large multinationals come into direct conflict with the environmental managers who represent charity organizations. No matter the course, it’s always for the good of the natural ecosystem.

1

Introduction to Environmental Management

CONTENTS

1.1. INTRODUCTION

We live in a world surrounded by biotic (living)[1] and abiotic (non-living)[2] components. The complex aggregated system of biotic and abiotic components or factors that surrounds and influences the form and behavior of an organism are referred to as the environment. The word environment originates from the French preposition "environ," meaning "to surround, envelop or encase" (Figure 1.1).

Figure 1.1. *The environment defined as the natural world in which land on which all organisms (animals, plants) live.*

Source: https://www.pbslearningmedia.org/resource/ketknh.sci.lifesci.threelevels/threebiolevels/.

Significant changes to Earth can be observed with chronological progression. In addition to the natural evolution of the planet, human beings have also influenced the physical world through our day-to-day activities as we attempt to pursue economic gain and increased convenience (O'Riordan, 2014). The changes to the environment have had both positive and negative impacts on human beings. To mitigate the negative impacts on human health

1 Biotic components refer to the living components of an ecosystem and includes animals, plants, fungi, and micro-organisms such as bacteria. Biotic components can be further classified, based on their source of energy, into Producers, Consumers or Decomposers. Producers (plants) generate their own energy through photosynthesis, in the presence of sunlight and chlorophyll. Consumers include herbivores, carnivores, and omnivores. Decomposers include saprophytes (fungi and bacteria) feed on decaying organic matter, recycling the nutrients in the process.

2 Abiotic (non-living) components are also referred to as ecological factors. They are the physical and/or chemical factors that act on the biotic components at any part of their life. Abiotic components include light, air, water, soil, nutrients, etc. These factors vary from ecosystem to ecosystem.

and well-being, attempts are made to restore or reinstate the environment. The term "Environmental Management" is multi-faceted and is simplistically used to describe the process of trying to bring balance to the conditions which affect our surroundings, human societies, and development processes so that they will not have adverse effects on us. The perspective and definition of the term "environmental management" and inter-alia of "environment" has evolved over the last few decades from one that encompasses an exclusively "green" perspective to one that integrates "green" issues with social, cultural, and economic issues ("brown") issues (Kotze and Nel, 2009)[3]. According to Barrow (2005) environmental management could refer to a goal or vision that attempts to steer a process with the application of a set of tools seeking to establish new perspectives towards the environment and human societies.

It includes the process of defining/describing and monitoring environmental change, predicting what environmental changes may occur in the future, with aims of maximizing human benefit from these changes and minimizing environmental degradation resulting from human activities. Environmental management brings together multidisciplinary fields of expertise and roles where the main objective is implementing strategies that aim at conserving natural resources and reducing the negative impacts on the environment and human societies, health, and well-being. The combined effect of the involvement of different role-players or stakeholders can ideally result in a more rapid discovery and implementation of solutions to emerging environmental crises (O'Riordan, 2014).

1.2. THE NECESSITY OF ENVIRONMENT MANAGEMENT

Environmental management gained prominence and came to the forefront in the 20th Century, largely brought about with public awareness of significant environmental incidents[4] (Levine, 1982; Medvedev, 1990; Wiens et al., 1996).

3 Kotze & Nel (2009) argue that the term "environment" for the purposes of environmental management should be considered from a legal and scientific perspective. The "green" definition of environment refers to the biotic and abiotic components of the earth system. The "brown" perspective incorporates human beings, and hence social issues, as an integral part of the system. Social issues may include the economic imbalances amongst different groups in the population, varying challenges related to individuals or population groups such as poverty, disease, crime, unemployment, and environmental justice.

4 Some of the most prominent large environmental incidents of the 20th Century include: The Love Canal Tragedy in the USA, Bhopal Disaster in India (December, 1984), OK Tedi environmental disaster in Papua New Guinea (1984), Chernobyl Nuclear Power Plant disaster in Ukraine (1986), Exxon Valdez oil spill in the Prince William Sound, Alaska (March 1989), Deepwater Horizon oil spill in the Gulf of Mexico (April to July 2010).

These disasters show the negative impact that human beings have had on the environment. They also reflect how intricately interwoven we are with the environment and the disproportionate impact of environmental disasters on poorer and marginalized communities. This increased awareness of our role in creating environmental disasters, which in turn has a negative impact on us has led to the urgency in the need to protect the environment. The concept of environmental management has evolved over time, as a result of increased awareness, understanding, activism, and legislative precedent. Major shifts have occurred in the adoption of a more goal-driven, consultative, pro-active or preventative approach to environmental management.

The world population increased from approximately 3 billion in 1960 to approximately 8 billion in 2017 (World Bank, 2020) This significant increase in the human population has led to many negative impacts on the environment such as increased deforestation, waste production, and pollution. Pollution rates have increased immensely with more industries being developed all over the world as humans seek to better their lives by manufacturing products which will make their lives more convenient. Economic greed has led to large mines mining activities with limited restorations of open pits or remediation of contaminated ground and surface water after the mines are abandoned (Russo, 2008).

The importance of protecting the environment cannot be overstated as we are reliant on the environment for our well-being. Without an environment that is suitable for our health and well-being, we will perish (Figure 1.2).

Figure 1.2: Abandoned house at Love Canal (USA).

Source: https://buffalonews.com/2018/08/04/gallery11053/.

1.3. RESPONSIBILITY OF ENVIRONMENTAL MANAGEMENT

Who is really responsible for protecting our environment?

The answer to this question lies deep down inside us. The environment belongs to us. Human society is inseparable from the environment. Whether we like it or not, we are all responsible for the preservation and protection of our environment. This responsibility cannot be outsourced to other stakeholders, including the government. The government is made up of only a limited number of elected representatives, who represent the interest of constituents. The primary role played by the government in environmental management is to legislate environmental laws and policies, facilitate reaching of consensus on critical environmental needs and the next course of action, and budget for the environmental needs in terms of policy implementation as well as enforcement are met. Decisions made by the government cannot be effective unless we decide to implement them. For the decisions are for us and by us and for the betterment of our lives.

As global citizens (Lee, 1993) we should and must take responsibility for managing our environment. Enough of solely delegating the responsibility to the government, to the nongovernmental organizations (NGOs), to our leaders in the society, to our religious leaders. Our environment belongs to us, and anything that concerns it should concern us too, as we form an integral part of the environment (Russo, 2008).

1.4. THE ROLE OF ENVIRONMENTAL MANAGERS

The most critical role of environmental management is to ensure that the environment in which all biota reside and depend is secure for their survival. Environmental managers play a critical role in the process. The role played by an environmental manager is complex and requires a diverse set of skills. Barrow (2005) states that "the environmental manager's job is to study and control processes to try and reach particular objectives." The concept of objectives in environmental management is important and serves as a golden thread to define the role played by environmental manager's *viz.* ensuring that the environment is safe for the health and well-being of all biota in balanced with the needs for development.

Environmental managers facilitate and implement the environmental awareness, consultation, and education campaigns to educate the public on techniques to understand the importance of a civil participation in reducing pollution rates, mitigate negative impacts from day-to-day activities on the environment. Thus begins the advent of the global citizen-informed about their surroundings, the role they play in their surroundings, the impacts that they have on their surroundings and vice versa and importantly armed with the skills in deciding and influencing the trajectory.

Environmental managers additionally play a key role in creating awareness of negative impacts from industrial processes, and work with other disciplines to mitigate, monitor, and manage negative impacts such as the downstream effects of waste generation (O'Riordan, 2014). Piasecki (1995) and Piasecki, Fletcher and Mendelson (1999) highlight the difference in the role played by internal environmental managers in the corporate/large business industries as compared to those in the private sector.

Managing of the environment introduces the concept of a systems approach, another skill that the modern-day environmental manager is required to understand and implement. An evolution of the definition of environmental management to one that is more goal-oriented facilitates the process to include the development of environmental strategies at local, regional, and global levels. Environmental managers design and integrate core strategies and processes to ensure that the environment is maintained at its peak performance at the current stage, but also catering for future predictions.

1.5. ENVIRONMENTAL MANAGEMENT FRAMEWORK (EMF)

An Environmental Management Framework (EMF) provides information that guides stakeholders on making the best decision on how to best solve the environmental challenges they are facing without causing further negative impacts. An EMF establishes clear procedures and methodologies for environmental planning, review, approval, and project implementation and monitoring. It includes information on whether the piece of land they plan to develop on is suitable for use, and if it is, what the impacts of the development process will be. Essentially, an EMF can be considered as a document that specifies what environmental limits should be included in spatial plans for development. The process of developing an EMF is important in identifying the roles and responsibilities of different stakeholders and also determining institutional arrangements that may be required such as training, capacity building, and technical expertise.

Several steps are involved in the development and implementation of an EMF. These generally include the following:

- Identify the project goals and the critical success factors.
- Identify stakeholders and options for management systems.
- Determine a sound understanding of the current environment. This involves a thorough study of the environment to evaluate its strengths and weaknesses.
- Determine potential impacts and prioritize solutions. An analysis is undertaken on what the extent of the impact of the proposed project will be in that environment. Once the evaluation process is complete, the stakeholders are consulted on the appropriate measures to be taken. They are informed on whether to proceed with the project or alternatives for the siting of the project are to be considered, should the environmental risks outweigh other factors.
- Decision-making on whether the impacts of the strategy are acceptable or not.
- Implementation of the agreed necessary actions with the assignment of agreed roles and responsibilities.
- Monitoring.

EMFs are successfully used for environmental projects, for the protection and restoration of natural resources. Sensitive protected areas, such as wetlands, must be considered to positively contribute to the society and not a hindrance to the development of the society. Community support is essential for the protection of environmentally sensitive areas.

The entire process of an environmental evaluation of an area environment and its designation as a sensitive matter is a long and resource-intensive procedure. It requires thorough scientific assessments.

1.6. ENVIRONMENTAL DATA: COLLECTION, MANAGEMENT, AND VISUALIZATION

The process of creating, collecting, and using data is resource-intensive and extremely valuable to an organization. The same controls that apply to other sectors in terms of data management also apply to data collected for environmental management. Environmental data collected must be reliable, requiring the proper management in order to avoid accidental loss or inadvertently providing access to unauthorized users and avoiding its alteration. What is meant by *proper management*? It is the process of ensuring data integrity from data collection until the conversion of the data to information. Data governance is a critical part of this process, whereby policies and best practices are developed and implemented.

Data collected for environmental management is obtained from different sources over varying durations, depending on the data objectives. An organization needs to centralize its information collected from different sources. Data management is the process of obtaining, validating, storing, safeguarding and processing vital data to provide accessibility, security and reliability to users.

As a best practice data management technique, data centralization by organizations like NGOs, government divisions and environmental consultancies ensures that relevant insights are obtained through data collation and that these insights can be successfully communicated to stakeholders and used for decision-making purposes (Russo, 2008). It relies on a centralized database where data is gathered, stored and maintained in a single location, but can still be accessed from multiple points.

With data collection, it is an added advantage for an organization to be able to track its progress by viewing its trends through the use of dashboards and other platforms. Dashboards enhance the visual representation of information making its interpretation and understanding quite easy and simple. However, visualisations are not restricted to the use of dashboards.

Still on best practice, an organization should ensure that they backup their information in a very secure location away from their current location and have an immediate recovery plan already in place to avoid loss of data in case of accidental erasure or disasters such as fires. Data governance and data security best protocol also prescribes using data archiving so that some data is not readily available. The planning part of data management is therefore important as it requires plans to be developed to discern which data should be readily available. There is also an argument to be made for not having all data centralized as this reduces the chances of losing data due to theft or changes to data from unauthorized access. A data back-up and recovery plan is important to ensure that data is maintained as it is a valuable asset to a company.

Data collected does contain some errors observed. It is up to the organization to ensure that their collected data is always clean. Cleaning data involves correcting all the misspelling errors, abbreviation errors, getting rid of all duplicate copies of data, and ensuring that documents are formatted correctly. Such practice ensures that the reporting process is completed within the shortest time possible (Russo, 2008).

One of the main reasons for having data errors is the use of manual entry procedure. With each manual entry of data, errors can be introduced. Organizations are supposed to update their entry techniques from manual entry of data to electronic data entry techniques. With electronic data entry, the chances of having errors are very minimal.

Without factual data, it's really difficult to run an organization. Hence, data governance needs to be the first thing to be considered before anything else. The access of data needs to be monitored for not every staff should be allowed to access information. Passwords need to be used in appropriate places and data being transmitted encrypted to avoid it being tapped during transmission.

Spatial information systems refer to computer structures working on data signifying spatial information in a broad sense: they utilize data found on, beneath or above earth. The systems naturally share multiple common properties with conventional database systems, but have recently given rise to additional spatial-tied systems known as spatial databases. In environmental sciences, they give experts the notable ability to consider fresh data sets related to newly gained and accurate data, plus to allow improved integration between diverse data sets. Furthermore, different Online resources with access to spatial data or maps can be found on the Web.

Additionally, Machine Learning methods originating from artificial intelligence are currently used in different aspects of environmental sciences. Due to their potent nonlinear modeling abilities, machine learning procedures today are applied in cable data processing, general circulatory models (GCM), air quality forecasting, , analysis and environmental data modeling, ecological modeling, including weather and climate predictions. (Russo, 2008) (Figure 1.3).

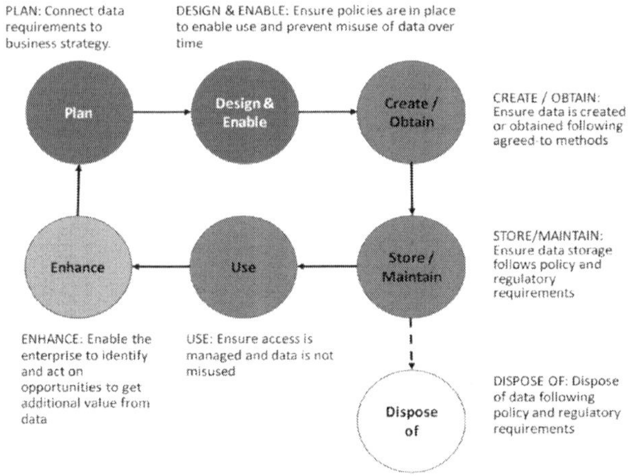

(Caption 1.3: Generic data governance and data lifecycle (Sebastian-Coleman, 2018)

Figure 1.10: Generic data governance and data lifecycle (Sebastian-Coleman, 2018).

1.7. IMPROVING ENVIRONMENTAL MANAGEMENT

Regulating environmental management requirements is quite a hard task for the government. Governments are not only tasked with the job of developing legislation, but also with the implementation and enforcement of the legislation. Negligence, lack of awareness, corporate greed proliferates in organizations and civil society, further complicating matters. Most organizations are not that willing to observe the proper ways of protecting the environment. They carelessly emit waste materials all over the surroundings exposing the environment to degradation. According to the World Bank, global waste will increase by a staggering 70% over the coming 30 years, standing at around 3.40 billion tonnes produced annually.

Furthermore, the involvement of civil society organizations in environmental matters is fundamental in helping governments to achieve more effective, democratic and viable agreements and decisions concerning environmental management. Equally, civil society plays a vital role in providing alternative measures for waste management and natural resource conservation initiatives. Some governments have implemented policies to try and restrict the emission of waste materials by industries. Where legislation is put into place and strictly enforced, polluting industries are rightfully required to comply with environmental management legislation. Industries may also "do the right thing" for purposes of accountability to shareholders, or as part of their corporate responsibility. (O'Riordan, 2014; Piasecki, 1995; Barrow, 2005).

Deduction of higher rates of taxes depending on the amount of pollution caused by a certain organization is another measure which has been put in place by some governments. The government calculates the amount of pollution brought by the so far emitted wastes and imposes the tax on the total earnings of the organization. This measure disciplines an organization's emission rates of wastes, for the more the deduction of taxes, the higher the monetary loss to a company. Some industries which emit poisonous wastes and are located closer to the residences end up being closed under the order of the government or relocated to locations far away from residences (O'Riordan, 2014).

Subsidizing the manufacturing processes of different products has been a great step in sound environmental management. Subsidy simply means the amount paid to industries by the government to ensure that they release less harmful wastes to the environment. The amount of subsidy needs to be higher than the total cost that a certain organization incurs in the general cleaning of the wastes that are released to the environment.

The use of tradable permits such as carbon trade permits is another way of improving management of the environment. Industries buy permits for a designated duration from the government or the society. The permits allow for the release of less harmful wastes for a specified duration. Once the duration is complete, the industry is not permitted to release anymore wastes. The permits can also be sold to the industries which obtain raw materials for manufacture of different products from the environment, such as mining and forestry. The industries are only allowed to mine a certain piece of land for certain duration of time and to certain levels. Once their duration is complete or the agreed level has been reached, the mining process is altered. The price of tradable permits is substantial because a certain amount of pollution is still allowed. Tradable permits serve as a useful option to promote the relationship amongst the government, civil society, and the industries as it is a means for everybody to be present at the negotiating table and requires consensus amongst the stakeholders (O'Riordan, 2014).

The use of deposit funds is one of the ways being used by the government to improve proper management of the environment. Products which are known to pollute the environment in one way or the other are sold at a higher price to try and discourage users from their purchase. The higher price is later refunded to the user once they commit to return the storage containers after use to a certain organization for proper disposal. Only a percentage of the initial prices is refunded to the user hence the name deposit funds, for during purchase, it's as if they deposited some money which would be given back to them once they fulfill their obligation of proper disposal of the containers, that is, return the container to the appropriate organization. The containers are mostly recycled for other uses since most of them are non-biodegradable.

The involvement of the public in the measures for improving the management of the environment is a great step to take. Both the public sector, private sector, and the government at large are united. Their unity does have an added advantage for the conclusions they all end up with are adhered to strictly and willingly by everyone without any objections. With everyone being involved in the conservation of the environment, the chances of ever facing environmental challenges are very minimal (O'Riordan, 2014).

Occupational/workplace exposure laws allow for the implementation of a safe working environment for employees. Negligence or non-compliance leads to a liability for corporations. Several countries have enacted workplace compensation for employees who have been injured as a result of exposure to toxic and harmful chemicals or unsafe working conditions, enabling employees to bring forth lawsuits against their employer. The lawsuit is not necessarily for compensation, but it acts as a motivation to workers rights (O'Riordan, 2014).

1.8. POSITIVE IMPACTS OF ENVIRONMENTAL MANAGEMENT

The proper management of the environment allows us to reap many positive results. Our day to day activities are made a bit easier for we are operating in an environment not harmful to our health and well-being, nor a physical environment so degraded that the plants and animals we rely on for our survival struggle for survival.

Symbiotic relationships are key to maintaining natural systems. People need clean air as we are reliant on the respiration for survival. Conserving our environment enhances clean air. The more trees we plant and preserve, the more we increase the amount of oxygen available for our survival. Trees take up carbon dioxide and utilize it during photosynthesis, producing oxygen as a by-product. Oxygen is then breathed in by human beings and they later breathe out carbon dioxide. Other creatures too do depend on either oxygen or carbon dioxide for their survival. Humans must ensure that they plant three more trees for each one tree which has been cut to prevent deforestation, this is because trees take long to grow and the only way to recompense for the one that has been cut is by planting 3 others. Furthermore, the chances of a tree growing to maturity are greater when more of them are planted at the same time. (Kotze, et al., 2009)

Ground vegetation also serves to improve air quality as they cover dust particles, thereby reducing the amount of particulate matter in the atmosphere. Plants also utilize other types of gases. Plants also play a critical role in the nitrogen cycle. Nitrogen in the atmosphere gets converted to nitrates by denitrifying bacteria in the soil. Plants consume nitrates through the groundwater manufacture of nucleotides and amino acids. Other gases like hydrogen are used by plants to enhance their general health. Generally, proper preservation of the environment improves the quality of the air we all receive as living organisms, both plants, and animals (Agarwal, 2005).

Sound environmental management helps reduce the biggest issue facing our planet, and that is, climate change. Temperatures have been increasing at high rates over the past few years. While the planet has seen several episodes of heating and cooling over the eras, anthropogenic factors have caused a steep increase in temperatures since the Industrial Revolution. Yes, the Industrial Revolution brought with it benefits such as improved living conditions, increased overall wealth. However, these improvements in living conditions have come at the long-term cost to the environment and to the survival of humans. Increases in temperature have been the knock-on effect of rapid and continuous development of industries. Humans have an insatiable desire to consume resulting in an ever-increasing number of industries to meet the demands of a steeply growing population. Gases released from industries end up damaging the ozone layer which protects the earth from direct sunlight that can be destructive to our atmosphere. Accumulation of the heat is causing a situation commonly referred to as heat island situation, for it is as if we are on an island of only high-temperature levels. Planting of trees does for trees tend to reduce the gases which accumulate forming ozone layer. Gases such as carbon dioxide are used by plants in the photosynthesis process, so the more we plant trees, the more the gas level drops making the ozone layer to be thinner, hence the temperature reductions. The main disadvantage of increases in temperature is that more energy will be used in cooling our surroundings. The planting of trees helps to increase the aeration rates of our surroundings, leading to the use of less energy for cooling and thereby resulting in savings.

We all need quality water. Preservation of the environment increases the quality of water resources. Trees and vegetation are an integral part of the water cycle through the process of evapotranspiration. Evapotranspiration involves plants absorbing water from the soil during the night and releasing it to the atmosphere during the day due to the high temperatures experienced during the day. The water vapor transpired by the leaves reaches a saturation point in clouds. The rain that subsequently falls reaches water bodies such as rivers and lakes. Vegetation ensures that soil remains anchored in place, minimizing surface runoff. Vegetation slows the rate at which runoff water flows, increasing the duration of stagnation in one place, and hence providing a sufficient time-lapse for the water to infiltrate into the ground. In cities, which have been covered by restrictive or impervious surfaces (such as concrete pavings), drainage ditches are developed to drain stormwater. Development of stormwater drains or ditches reduces the imperviousness of the ground hence a reduction of the flooding rates in the surroundings. Industries need to clean their waste materials before releasing them to the nearest water bodies (Figure 1.4).

Figure 1.4: Clean water helps preserve the habitat of the aquatic animals.

Source: https://www.worldwildlife.org/initiatives/oceans.

Buffers are engineered to pipes draining runoff water from different urban cities into water bodies to ensure that water is free of all solid waste material before entering water bodies. If possible, the buffers should be constructed in such a way that cleaning chemicals can be incorporated into the pipes for the water to be cleaned before being drained into the water bodies. High vegetation cover increases the rate of ground infiltration of precipitation. Soil deposited into water bodies causes significant problems over time. These include increased risk of flooding due to raised water levels, eutrophication due to increased nutrients that are slowly released from the soil and stagnation of water (Agarwal, 2005).

We are part of a whole, and the whole is made up of several ecosystems and a full range of biodiversity. We rely on this diversity for our very survival. The protection of natural habitats is key to protecting biodiversity. We all need to preserve the environment so as to protect different species. These habitats serve as homes, mating, and breeding grounds for various species. Unpolluted water equates to aquatic species having a clean habitat. The conservation of plants, vegetation, and our natural surroundings, means the many habitats to many species are secure for their survival. We need insects for them to pollinate our plants which in turn provide us with our daily means. We need to preserve plants for most of them have medicinal benefits to us. We need to preserve the aquatic environment for we depend on aquatic species for so many uses. We need to preserve natural habitats for most of them serve as tourist attraction sites providing a means of income. We also need to plant more trees and preserve the existing ones for serving as windbreakers, protecting our homes and other infrastructure from strong winds. We too as human beings need natural surroundings for recreational purposes, such as hiking, water sports, and other fun outdoor activities that contribute to our sense of well-being. It's high time we value our biodiversity and preserve the environment at large for it offers a plethora of benefits to us (Agarwal, 2005).

Can a more beautiful environment create a sense of "green shaming" where we get to act in a "greener," more environmentally friendly manner, and abandon our recklessness towards the environment?

The preservation of the environment does encourage us to act greener. The beauty of a well-preserved environment does promote ecotourism. Ecotourism motivates us to innovate more ways of greening our environment. With ecotourism, most people tend to opt to implement ways of performing different activities in a less harmful way to their surroundings.

1.8.1. Challenges of Preserving the Environment

The development of open space environments more so in urban areas is a great responsibility. The ones responsible for the planning of such projects do face various challenges before they actually come to the stage of implementation of their project or during the implementation process.

The availability of enough space for the setup of open surroundings within the urban location is challenging. Having a few large open surroundings with natural vegetation does not assure that the whole city will be fairly aerated. It does require several small-sized parks to be set around the city for it to perfectly benefit from their natural characteristics. Most of the setup of the park in urban areas is used for recreational purposes limiting the growth of natural vegetation yet the main aim of their setup was to promote the growth of natural vegetation (Uberoi, 2004).

The parks are filled with pavements which promote imperviousness reducing the rates of infiltration in the surroundings. This leads to increased rates of surface runoff which is drained into water bodies yet in most regions, drained water to water bodies is not really connected to any buffers.

Any attempts to further implement the parks in urban centers to be used for more growth of natural vegetation do bring unwanted conflict. Most of the residents prefer the parks to be used for recreational purposes yet the environmentalists' main aim is to perfect the parks for natural purposes.

The main challenge is that preserving land for improvement of biodiversity means having to sacrifice other benefits such as the use of land for recreational purposes (Uberoi, 2004).

1.9. NEGATIVE EFFECTS CAUSED BY ENVIRONMENT MANAGEMENT

The setting up of open spaces in urban areas does cause some negative effects to their surroundings. Road routes end up being long for most of the parks are located in between the cities. The number of road routes is limited to such that vehicles have a limited number of ways to navigate the city from one place to another. The fewer the road routes, the higher the rates of traffic jams in the urban centers. People end up taking longer periods to navigate from one place to another within the same routes.

Most of the parks are used for recreational purposes. The main reason for setting them up is to preserve the environment yet they end up causing more pollution. Where more people are gathered together, the chances of pollution are high. This is because people tend to dispose of their wastes more so the plastic waste anywhere around their surroundings. More noise is caused around the city by children having fun in the parks among other disadvantages (Cassar, 2013).

The potential indicators of a well-preserved environment are a great motivation to environmentalists. A surrounding with perfect air aeration of high quality, reduced greenhouse effects, and well-balanced biodiversity is a good indicator of a well-managed environment.

1.9.1. Human Impact on the Environment

Human beings have immensely contributed to the ever-changing conditions of the environment. Their impacts are of different types (Figure 1.5).

Figure 1.5: Overpopulation is one of the ways in which we have altered the environment.

Source: http://theconversation.com/overpopulation-and-the-environment-three-ideas-on-how-to-discuss-it-in-a-sensitive-way-98361.

The number of people on our planet is almost exceeding the limit it can accommodate. Improved health care facilities have increased the number of people by reducing the mortality rate worldwide. Despite the introduction of family planning options, the world population is still growing. As the world population is increasing, more land is being cleared for the setting up of settlements. The current generation is innovating all kinds of materials to make their life easier. The manufacturing processes do require the establishment of industries which occupy space hence deforestation of our surroundings. The dependency on fossil fuels such as coal, charcoal, and petroleum do have more negative impacts on the environment. As more land is being cleared, gases such as carbon dioxide are accumulating in our atmosphere at higher rates since there is no vegetation to clear up the air in our surroundings. The burning of charcoal for fossil fuels is leaving more and more land deforested. Fossil fuels produce harmful gases to the environment such as carbon dioxide. Generally, the more the population, the higher the rate of environment degradation (Cassar, 2013).

Human behavior of disposing wastes in the right way has completely vanished. People no longer care about their surroundings. Such situations have become more and more serious to an extent that industries are releasing poisonous gases to our environment without bothering to at least recycle or clean the gases before disposing them off to the atmosphere. Rivers are being polluted due to disposal of untreated waste to the main streams in our surroundings. Environmental organizations are no longer strict on the proper environment for setting up certain industries. They are authorizing the setting up of industries releasing harmful waste next to human residences. The soils are being polluted too due to excess use of pesticides and insecticides by farmers (Figure 1.6).

Figure 1.6: Pollution is everywhere in our current world. From candy storage papers in our streets to the large pits in our cities filled with all kinds of waste material.

Source: https://www.nationalgeographic.com/magazine/2018/06/plastic-planet-waste-pollution-trash-crisis/.

As more forest land is being cleared for different activities, more rates of carbon dioxide are accumulating in our atmosphere. People have completely neglected their role of carrying our reforestation practices in deforested areas. Fossil fuels are being used at higher rates by humans. The combustion of fossil fuels does increase the accumulation rates of carbon dioxide in the atmosphere. With more accumulation of such gases, the globe's temperature is increasing. With the increase in temperature, the ice usually accumulated at the arctic area of both the north and the South Pole of our globe is melting at higher rates. It is estimated that in the next few centuries, the sea level will rise to higher levels which might cause flooding effects to our surroundings (Cassar, 2013) (Figure 1.7).

Figure 1.7: Global warming is a major human-induced environmental impact.

Source: https://www.reusethisbag.com/articles/10-human-causes-of-global-warming/.

Weather patterns are changing at higher rates in our globe. Some areas which experienced longer winter seasons and shorter summer seasons are experiencing the opposite to their normal weather patterns. Some areas which had enough supply of water are suffering from droughts. Agricultural patterns, which have been passed on for generations, have been interrupted due to a changing climate. The chances of experiencing hurricanes more frequently are rising and this may leave many people homeless. Flooding rates are increasing in the lowland areas due to the climate change occurring globally. Most parts of the globe will experience prolonged periods of drought with deserted areas having very high degrees of temperatures. All these effects on the climate have been caused mostly by the industries being set up by human beings for the perfection of their lives. The ever-advancing technology does have some contribution in the changing climate.

Environmental challenges call for innovative solutions and solving problems is what humans thrive on. However, not all solutions result in the desired outcomes. Take the use of genetically modified organisms (GMOs) to solve the problem of growing food in water-scarce areas. GMO plants, such as canola, alfafa, apples, cotton and potatoe structures have been modified to their deoxyribonucleic acid (DNA) so that they can be grown without the need of much watering. They are grown in countries such as USA 70.9 Mha, Brazil 44.2 Mha, Argentina 24.5 Mha, India 11.6 Mha, and Canada 11 Mha.

The use of herbicides to get rid of weeds has been of great use though a large percentage of weeds are getting resistant to the herbicides. Resistance to particular herbicides is among the key traits of genetically modified organisms, also known as GMOs. Farmers may spray the entire field, but just the weeds would die. Nevertheless, weeds adapt and may evolve just like every other living species. As such, they can develop some resistance to herbicides across time. For instance, nowadays weeds that are resilient to glyphosate compound have emerged and are often known as 'super weeds'.

This leaves farmers with only one option of having to till the land. Tilling the land does leave the soil less fertile though the use of fertilizers helps boost the nutrient content of the soil. The main disadvantage of the use of fertilizers is that they pollute the soil in one way or the other. The more we as humans opt for genetically modified plants and organisms, the more we end up causing more harm to the environment (Cassar, 2013) (Figure 1.8).

Figure 1.8: Significant emissions of carbon dioxide by industries have contributed to climate change.

Source: https://www.nationalgeographic.com/environment/global-warming/pollution/.

Once the carbon dioxide comes into contact with the salty ocean water, carbonic acid is formed. Carbonic acid leads to the destruction of coral reefs through coral bleaching. Coral reefs act as a primary habitat for many of the aquatic animals, making them some of the most diverse and valuable ecosystems on the planet[5]. The increase in the amount of carbon dioxide gas in our surroundings is all caused by the mismanagement of the environment by human beings.

Aquatic life in all water bodies including rivers is at a higher risk. During the rainy seasons, the excessive amounts of fertilizers we apply to our crops are being carried away into the water bodies. As we all know, fertilizers contain chemicals which boost the growth of plants.

5 According to the National Oceanic and Atmospheric Administration coral reefs support about "4000 species of fish, 800 pieces of hard corals and hundreds of other species." As far as marine real estate is concerned, very few places compares in terms of value in supporting marine ecosystems compared to this.

Figure 1.9: Land clearing.

Source: https://www.pinterest.com/pin/684688568064049011/.

As fertilizer continues to accumulate in our water bodies, more water hyacinth keeps on covering the water surfaces leading to lowered aeration rates. Aquatic species end up dying due to lack of enough oxygen. Garbage is also being carried into our water bodies. Most of the wastes are not soluble and they end up being deposited in the lower levels of our water bodies. Aquatic species end up feeding on these wastes considering them to be food which in turn leads to the suffocation of aquatic animals (Cassar, 2013) (Figure 1.9).

People are also relying on charcoal obtained from trees hence increase in the tracks of land which are left deforested. The more deforested our surroundings are, the more we have to suffer due to insufficient supply of oxygen for plants are our main source of oxygen. With fewer trees, the greenhouse effect, which is the accumulation of harmful gases in the lower levels of our atmosphere, is increasing (Cassar, 2013).

Combustion of fossil fuels such as coal, charcoal among others has been known to release poisonous gases to the atmosphere. The gases rise up the atmosphere to the concentration level where clouds are formed. The rain-bearing clouds formed end up having a mixture of the gases. Once it rains, the rainwater ends up being acidic. Accumulation of the acidic rainwater in water bodies ends up acidifying the water which in turn depletes oxygen level. The depletion of oxygen levels leads to mass deaths of aquatic life. Infiltration of the acid rain into the ground depletes the soil of its important nutrients and soil living organisms end up dying. As plants absorb water from the acidic rain, they are scorched and they end up drying. Acid rain does affect buildings for it scorches iron sheets used in buildings among other damages.

The ozone layer has been serving us for ages saving us from dangerous effects. With the emission of poisonous chemical gases to the atmosphere, the dear layer has been growing thinner with each passing decade. As the layer became thinner, more dangerous ultraviolet rays were able to reach the earth's surface. The ultraviolet rays are of immense distraction to human skin, plants among other species on the earth's surface. The depletion of the ozone layer has brought about incurable skin diseases, death of plant life among other negative impacts (Russo, 2008).

As human beings, we need to pull up our socks and really put into consideration the preservation of our environment. Mother Nature is a very powerful force, and it's high time we treat her right before her revenge mechanisms get too intense for us to persevere.

1.10. ENVIRONMENT MANAGEMENT CONCEPT

The main concept of environmental management is to ensure that the environment is perfectly preserved. By preserving the environment, human life is sure to be protected, species will be able to survive in our surroundings without the fear that they will become extinct among other benefits.

1.10.1. Scope of Environmental Management

Environmental managements basically deal with ensuring that the relationship between the social and economic development of human life are perfectly balanced and that our surroundings environment is perfectly well for such development activities.

1.10.2. Aspects of Environmental Management

Environmentalists have put into consideration several ways in which to improve the preservation of our environment.

Creating awareness to the public on the importance of preserving the environment is one of the efforts put in place by environmental organizations. Advertisements on the ways of preserving the environment are being aired on our radio stations, television channels, billboards among other places. Seminars, workshops, and tree planting days are being arranged every now and then to try and reduce the pollution rates of our surroundings.

Firms are being educated on better ways of managing the resources they use for manufacture of different products. With proper management of resources, it ensures that the environment is perfectly preserved for there is an extreme to what extent they should obtain a certain resource avoiding destruction of our surroundings.

The release of harmful products into the environment is being controlled by environmental organizations. The rates of pollution are also being controlled to ensure that we at least try to make our surroundings clean and fit for our survival (Russo, 2008).

Organizations responsible for helping in the preservation of the environment are being assessed regularly to ensure that they perform their roles to their level best.

1.10.3. Environmental Management Plan (EMP)

Every environmental management organization should have a well-defined plan of the measures they have put in place in order to manage our environments in the most effective way. An environment management plan could also be formulated by industries as they try to figure out what environment preservation techniques they should put in place to ensure they take part in making their surroundings a better place. An environment management plan should follow a certain procedure for it to be termed as a perfect plan.

The first step in coming up with a perfect plan is outlining all the goals and the policies which govern the operations of a certain firm or organization. Identifying the most crucial environment regulations is a very vital step because, before any operations commence, an organization should be aware of the limits they should not cross when it comes to issues to do with the environment. If any tax fees are to be charged depending on the environmental regulations, they should be calculated early in advance to ensure that the goals are in line with the deductible fees. All the goals an organization comes up with should have a specific deadline of when they should be met. As an organization, it is advisable to have an in made environment policy apart from the ones set up by the government. It should be more of a motto for the company in terms of preserving the environment (Russo, 2008).

Once the planning part is complete, it's advisable that your agreements be documented in a perfect manual. The manual should be arranged in a specific order that shows the organization is serious about the plan. There rules which should be followed for a certain plan to be considered presentable to anyone, they should be strictly adhered to for this plan to be represented on various platforms. If any training activities need to be conducted to aid in final compilation of the plan, they should be well arranged to avoid any inconveniences.

Once the documentation of the plan is complete, it's time to put your plan into action. An organization should divide its specific goals amongst its staff. Performance expectations should be perfectly outlined before assigning goals. Regular progress checks should be conducted and communicated to the authorities for further scrutiny (Russo, 2008).

1.11. ENVIRONMENT MANAGEMENT SYSTEM

The duties for ensuring that the environment is preserved are growing with increase in the negative effects affecting our surrounding. Technology experts have linked up with environmental management organizations so as to come up with a system that will help in managing the environment. In simple terms, an environment management system is a collection of different entities which ensures that all the policies of a certain organization are adhered to and that all its plans are implemented.

The main objectives of setting up an environment management system are to ensure that the activities of an organization are performed in an orderly manner. In situations where references are to be made, an environment management system has perfectly documented everything that has to do with the entire organization. It also ensures that all finances are managed properly and there are no fraud issues within the organization, for with electronic management of information, it is hard to mismanage any type of information.

The cost of setting up an environment management system is quite high. This is because electronic components to be used in the new system are expensive to purchase, the training of staff for them to be able to use them (Uberoi, 2004).

An environment management system does have its main roles across all setups despite the type of organization it is being implemented in.

One, the system should be able to help plan all the organization activities, that is, it should organize a firm's information in an orderly manner. By organizing information in an orderly manner ensures that updated reports are produced regularly so as to help the organization in making its decisions. The reports do guide the organization in coming up with its regular plans on how to conduct different projects in an organization.

After the plans have been formulated, the organization should start acting in the plans. It is the work of a system to keep track of how the plans are being acted up (Uberoi, 2004).

Employees are required to update their progress on their respective projects in the system. The employer is to be able to monitor the execution of different projects without having to directly supervise.

Once a certain plan has been implemented, it is up to the system to help the authority in final evaluation. Any errors made during the implementation process are reviewed and corrected with the help of the system.

1.11.1. Benefits of an Environment Management System

An environment management system ensures that there is a central control of all activities in an organization. This is because all information concerning a certain organization is all available in one central location.

It is beneficial in the fact that it does help in reducing the rates of pollution. It measures the pollution rates both quantitatively and qualitatively making it easy for the government to calculate the total fee to charge for pollution of the environment.

The system helps boost the reputation of an organization. With the system, an organization is able to focus on its critical issues and to their level best. Finances are managed properly with the help of the system among other advantages.

1.11.2. Principles of Environment Management

The government has set up beliefs which govern how the public perceive that preservation of the environment. These beliefs are there to challenge people to forcefully ensure that their surroundings are clean or they will have to face the consequences.

One of the key principles of environmental management is the polluter pays principle. It reminds the public that, whoever polluted their surroundings in any way, is in for the consequences. If not anyone takes the responsibility of cleaning the environment once they pollute it, they are charged huge amounts of money to cater for the cleaning of their surroundings since they are the reason it is nor clean (Agarwal, 2005).

User pay principles are basically for consumers. It reminds them that they are responsible for boosting the health of their surroundings by disposing their used waste material properly. This is because they are the ones who purchased the materials, and used the contents in them therefore; it is up to them to dispose them well. Failure to proper disposal of used material, they are charged a fee to support proper disposal of used materials.

The other principle is the precautionary pay principle. It is a fee charged on individuals who are not precautious enough in the handling of various waste materials. It challenges people to be very careful in the way they dispose of their waste materials without risking their lives or the lives of people within their surroundings (Agarwal, 2005).

The principle of effectiveness and efficiency reminds people that they should consider preservation of the environment to be one among their highly ranked priorities. It reminds people that they should not wait until tomorrow to clean their surroundings by charging them some fee.

Principle of responsibility is a fee charged on people who are not responsible enough for the cleanliness of their surroundings. They are regularly reminded of the importance of being in clean environments.

The other principle is the principle of participation. We are all supposed to take part in cleaning our environment. In tree planting days, we should all show up for the day and be ready to participate. In community clean up days, we all should be at the forefront cleaning our environments. This principle charges individuals who do not participate in such activities some extra fee for failing to show up.

Last but not least, the principle of proportionality. It ensures that we all have equal contribution in cleaning up our surroundings. It charges some fee from those who live the responsibility of cleaning up their surroundings only to one person (Agarwal, 2005). All the discussed above principles have played a major role in preserving our environments.

1.12. INTEGRATED MANAGEMENT SYSTEM

The term environment is a really ambiguous word. It entails so many components which all need to be addressed. We need to discuss of an environment management system earlier which has been of great help so far. However, with so many components of the environment, each of the components does require a single management system to control the operations of its surroundings. The management of so many systems all at once tends to be very tiresome to the government.

The above-discussed challenge led to the introduction of a new way of managing multiple systems all at once. An integrated management system is a system that simultaneously manages multiple components of an organization. It can be thought of as analogous to a mature, superior tree with several small branches representing multiple management systems that deal with a specific component. In order to be fully functional, it is imperative that each component within an integrated management system is perfectly managed (O'Riordan, 2014) (Figure 1.10).

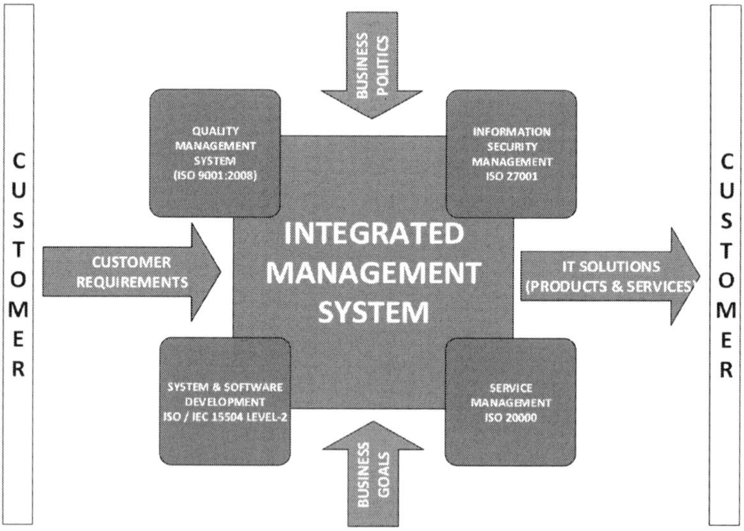

Figure 1.10: An illustration of integrated management system.

Source: https://www.researchgate.net/figure/INNOVA-Integrated-Management-System-Diagram-INNOVA-IT-Solutions-plans-applies-checks_fig4_263818007.

1.12.1. Benefits of an Integrated Management System

An integrated management system is basically a whole large system which approaches its subsystem as part of it and as a sum of its parts. It has benefitted many organizations in so many ways.

It eliminates the duplication of information in an entire organization. This is because information is managed from a central location hence no need of handling it from its entry location. Such activities relieve staff of their redundant activities for they are able to focus on the most critical tasks (O'Riordan, 2014).

Senior managers of an organization have a lot to do. The managerial staff should do their best to ensure that they do not take a certain t of the senior's quality time. With an integrated management system, the most important information is able to be retrieved in an orderly manner and presented to the senior for them to make their final decision.

Resources need to be management in the most effective way. With an integrated system, the planning of the quantity and quality of the resources required is well calculated and the duration it will take for the resources to be obtained in situations where they are located at distant locations. Such management of resources is achievable because all information of an organization is all located in a central location. With an integrated management system, the likelihood of having mismanagement of resources is minimal.

The cost of conducting different operations within an organization is reduced. This is because, with central management of all components of an organization, the total cost that an organization will incur in its operation is perfectly calculated and well planned for. It reduces cases to do with fraud for no one is able to lie about the cost of any operation (O'Riordan, 2014).

Organizations tend to have regular checkup examinations conducted on them either by the government or any other body. Because of the centralization of information in an integrated system, all the examination personnel have to do is visit only one point in the entire location for complete evaluation of the entire organization. With an organization whose system is not integrated, it would take regular checkup examination to be conducted for its evaluation to be complete, for each component will require to be evaluated thoroughly, and definitely, the chances of having multiple cases of duplication are high.

1.12.2. Limitations of Integrated Management System

Just like a coin, everything in this universe does have its positive and its negative side. For an integrated management system, the positives do outweigh the negatives. The drawbacks of integrated systems are not that intense.

With the integration of multiple systems, the chances of losing focus are very high. Integrated system involves the linking of systems which deal with different kinds of information. For example, the organization might decide to improve the performance of all their subsystems; this might confuse the staff since the implementation of a quality management system is different from the implementation of the energy management system. With this confusion, employees lose focus on what part they should really focus on. All this leads to mismanagement of the entire organization rendering the whole system useless and of great loss to the organization (O'Riordan, 2014).

Each system among the multiple managed systems does have objectives. The objectives of the quality management system are completely different from the objectives of energy management system, food safety management system, information management system among others. All the objectives do require different resources of different costs for them to be fully implemented. This leaves an organization with the responsibility of having to choose which objectives are critical. The elimination process of some objectives is a long and tiresome one. It might force the organization to fail meeting its annual resolution due to lack of enough resources.

The management of a single system seems to be easier than the management of an integration of multiple systems.

1.12.3. Components of an Integrated Management System

As discussed earlier, an integrated system is a main system which handles multiple systems. We need to understand what type of multiple systems it does combine for it to be termed as an integrated system.

Quality management system is one of the multiple systems managed by the integrated system. As its name suggests, its main role is to ensure that all the services delivered to customers or all the products manufactured or generally, all the activities of a certain organization are of high quality. It eliminates any chances of having any complaints filed against them because of offering low-quality services. It promotes the entire organization by boosting its reputation (Russo, 2008).

Environment management system is part of the multiple systems under an integrated system. As discussed earlier, it monitors and controls the operations of an organization within its environment. It ensures that there are no irregularities conducted within the environment of a certain organization.

The other subsystem is the energy management system. Its main role is to monitor the consumption of energy within its environment. It tries to ensure that the entire company does not mismanage the energy within its reach. It does also ensure that the type of energy being used is clean energy that does not end up polluting its surroundings. Many organizations are doing their best to make use of the emerging energy-saving techniques to power their operations.

Food safety management system, as its name suggests, deals with food issues. It governs the production of food from the farms by ensuring that there is sufficient labor for farm activities, enough funds are invested in the farm for different uses such as the purchase of herbicides, fertilizers among others. During harvest time, it is up to this system to ensure that farm produce is obtained in time, the portion for storage is properly stored for future use in perfect conditions, and the portion for manufacturing processes is transported to their respective processing units. It does handle all the finances dealing with food production, processing, and the profits (Russo, 2008).

Information security management system is a really crucial subsystem of the main integrated system. It ensures that all data that has to do with the entire organization is protected from unauthorized access by all means. It does also handle recovery plans by backing up the entire information in a different location to be used later in situations such as accidental loss of data either due erasure or disasters.

The discussed above components are all linked together in one way or the other in an integrated management system to ensure that they all operate uniformly. This is because they do depend on each in different ways. Their integration improves the performance of an entire organization (Russo, 2008).

1.13. ENVIRONMENTAL LEGISLATION

As discussed earlier, the environment is a very sacred resource to us all, for without the environment we are no more, the environment is us and we are the environment. Rules and regulations had been setup, to help in limiting the mismanagement of the environment. They help in reminding individuals that, it's up to us to take care of our mother nature, for it does take care of us too.

We need clean air for us to survive. Different species in our surroundings need clean air too for them to survive. A clean air legislation act has been implemented in various countries of the world to ensure that the surroundings have clean air. Industries found emitting poisonous gases to the environment are being charged huge amounts of money and some end up being shut down completely with no future of ever coming back into business for endangering the lives of people. Paper processing companies are given a rule of governing more plantation of trees since they are the ones responsible for the widely spread logging activities. All these regulations have been set to ensure that we breathe in clean air (Russo, 2008).

Poaching activities have led to near extinction of various species. Governments have set up laws and regulations to try and eliminate poaching activities. Any individual involved in poaching has to face very serious charges in a court of law. Such individuals are being fined huge amounts of money and imprisoned for many years. The shipping of parts from poached animals is being illegalized. Shipping companies are made responsible for any shipment of illegalized materials. Cargo planes have been banned from transporting illegal animal parts. The security in our parks has been tightened. Any corruption issues to do with poaching activities are being dealt with seriously. Other activities which are risking the lives of the almost extinct species are being banned. The fishing of the almost endangered aquatic species has been illegalized and serious measures have been put in place to deal with such cases (Russo, 2008).

It is said that water is life. Clean water act has been implemented to deal with individuals who are not willing to maintain their aquatic surroundings clean. Companies are being forced to completely clean any waste materials before disposing them off into the nearby water bodies for failure to do so; they have to face the strict consequences of either complete shutdown, or high charged fines or even both. Settlements in water catchment areas are being burned so as to ensure preservation of the sacred places, for they provide us with water. Farmers are being taught on proper uses of herbicides and fertilizers so as to avoid pollution of water bodies. The failure of farmers to use adequate amounts of farm inputs, they are being charged some fine to cater for the cleaning of water bodies.

An act has been implemented to ensure the preservation of the ozone layer. Burning of fossils fuels such as coal has been banned. Innovators have been given the chance to innovate energy-saving techniques which promote preservation of the environment such as the use of energy saving jiko (which makes use of small amounts of charcoal at a time). With less combustion of fossil fuels, there is less release of poisonous gases into the atmosphere hence the restore process of the ozone layer is boosted. Without the ozone layer, every living species in the entire continent will end up being distinct (Agarwal, 2005).

2

Environmental Issues

CONTENTS

2.1. INTRODUCTION

Currently, the world is facing a lot of environmental challenges. These challenges are either attributed to man or naturally caused. This chapter was prepared to provide a glimpse into some of these issues.

2.2. ATMOSPHERIC ISSUES

The earth's atmosphere has been continuously being damaged by human activities in different ways and fronts. The stability of the atmosphere is important to maintain a healthy environment for animals, plants, and humans. The earth has a protective layer called the ozone layer and without it, the survival of man and other life forms will be compromised. This layer overtime is being depleted by chlorofluorocarbons and other destructive chemicals such as bromofluorocarbons which release destructive free radical catalysts such as atomic bromine and chlorine, hydroxyl, nitrous oxide, and nitric oxide. The ozone layers' major function is to have an absorbent effect on the sun's ultraviolet radiation as it contains a higher concentration of ozone in comparison to other areas of the earth's atmosphere thus consequently there is the release of ultraviolet rays to the earth which are unabsorbed and dangerous to different life forms (Rangarajan, 2009) (Figure 2.1).

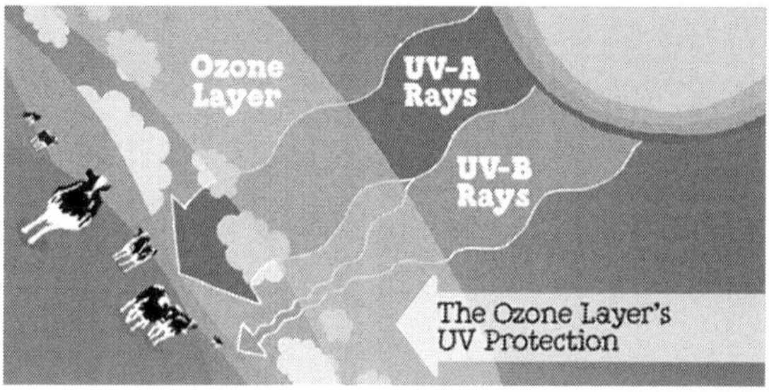

Figure 2.1: The ozone layer serves as a barrier from the sun's harmful UV rays.

Source: https://www.benjerry.com/whats-new/2014/ozone-layer-recovery.

2.2.1. Effects of Ozone Depletion and Climate Change to the Atmosphere

Since the late 1970s, ozone levels have decreased by about 4% per annum as a result of human activities. Scientists have been able to confirm this destruction with the use of satellites and it can be said to be far much worse than was anticipated. The resultant effect of this destruction is the growing numbers of cancer cases especially skin cancer, and other immune disorders, climate changes, and also changes in the plankton population which is food for marine animals. This has seen the banning of the use of chlorofluorocarbons in many countries which the aim of saving the planet[1].

Large quantities of carbon dioxide are released to the environment with the use of fossil fuel commonly referred to as our carbon footprints. We cannot only blame humans but also animals such as cattle which also release huge amounts of methane both of which trap the earth's heat and prevents back radiation to the atmosphere. The resultant effect is the contribution to the destabilization of the climate on earth and thus the greenhouse effect which is the warming trend. These changes may not give enough time for fragile ecological systems to survive. Sometimes widespread hunger leading to malnutrition occurs due to damage of crop production capacity in different agricultural zones (Rangarajan, 2009).

In 1992, the Kyoto protocol was formulated and established at a UN convention on climate change and it commits its member states to reduce greenhouse emissions to reduce and prevent global warming and is caused mainly by carbon dioxide. The Kyoto protocol created in Japan became effective in 2015 by the 192 member states. And even though 36 developed countries worked to have a reduction in their carbon emissions, the overall effect that was intended was not achieved as emissions increased by 32% globally between the year 1990 and 2010. There is another amendment after the second commitment in 2012 known as the Paris amendment and the current Doha amendment which was effective as of February 2020.

2.2.2. Policy Issues in the Protection of the Earth's Atmosphere-Ozone Repairing and Depletion

Strategic measures can be put in place and measures employed in one region similarly be done in another region. They are actions that have been and can be put in place to mitigate and counter this effect. They include banning the use of CFCs throughout the world and other products that destroy the ozone layer at the same time doing research on other substances that are less toxic

1 The Montreal Protocol on Substances that Deplete the Ozone Layer (is a multilateral environmental agreement that regulates the production and use of early 100 man-made chemicals referred to as ozone depleting substances (ODS).

to the environment. Environmentalists and policymakers can find ways to implement policies or create legislations that see the reduction of carbon emissions. The common steps that can be taken by different nations include reliance on renewable energy sources, use of nonpolluting energy sources by industrial plants, power plants and the transport sector and increasing efficiency in the use of energy sources.

2.2.3. Policy Opportunities in Relation to Climate Change

To control climate change cooperation among different countries to work towards reducing overdependence on fossil fuels is necessary this can be done by the use of renewable and clean energy sources. Banning of tire derived sources, reusing, and recycling used tires in an eco-friendly manner rather than burning them up (Rangarajan, 2009).

Stopping deforestation is another appropriate move as climate change is worsened by deforestation as trees take up carbon dioxide and lack of trees means an increase in circulating carbon emission. Instead of cutting trees, more trees should be planted to help preserve the environment and the atmosphere.

Other policy opportunities include strong response to the Berlin mandate which involves the creation of short, medium, and long-term goals for reducing greenhouse gas emissions and the agreement binding countries would be an incentive to industry to reduce carbon emissions.

With time, many countries have been able to make notable progress towards the elimination of chemicals that deplete the ozone layer. To repair the ozone layer and prevent depletion policies that can be put in place include strengthening of the Montreal Protocol which advocates for the removal of the use of methyl bromide and establishment of a fund especially by and for developing countries to implement the Montreal Protocol. Development of alternatives to ozone-depleting substances that are safe and also develop technologies that are low in cost that can safely eliminate this ozone-depleting chemical and finally the enforcement of legislation of cfcs and the control of the illegal trade of these cfcs (Rangarajan, 2009).

2.3. LAND SUSTAINABILITY ISSUES

Land is a natural resource that supplies a wide array of physical and socioeconomic goods and services which provide support to ecosystem sustainability, the wellbeing of humans, and finally livelihoods. This being the case there are threats to this role created by land degradation and desertification which consequently affect the fertility of this land and other benefits to the human race and society. In 2015, a report done by the economics of land initiative showed that a percentage of land 10–20% worldwide has been degraded which amounts to 6–12 million square kilometers which is almost a quarter of productive land on the earth. This affects around 3.2 billion people and as a result leads to water and food insecurity and finally poverty (Singh, 2005).

2.3.1. Causes of Land Degradation

Land degradation is an environmental problem worldwide. The major causes of land degradation are population pressure, urban development, cutting down trees, land pollution and improper waste management, unsustainable land management practices, and climatic changes.

2.3.2. Consequences of Land Degradation

For affected areas land degradation leads to an increase in vulnerability to climate change. This also increases a strain in land and water resources in addition to the pressure on resources which is a result of the ever-growing earth population which results in an increase in the demand for food especially those coming from animals which also leave behind a large carbon footprint. The data from the ELD initiative as of 2008 measures the consequence of land degradation in numbers. It measures annual global losses as a result of deforestation and degradation as 1.5–3.4 trillion Euros hence the need to stop deforestation and encourage reforestation and afforestation to ensure that available land continues to provide sufficient land-based ecosystem services even with the increasing population (Figure 2.2).

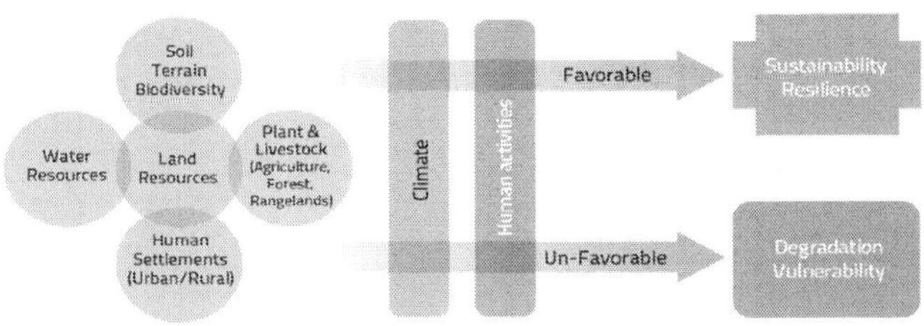

Figure 2.2: Sustainable land management (SLM).

Source: http://www.fao.org/land-water/land/sustainable-land-management/en/.

The goals of sustainable land management (SLM) include maintaining the functions of an ecosystem and services while concurrently supporting the wellbeing of the human population. SLM has eased in adaptability and has been encouraged throughout the world and can be used in the preservation of land. In 1992 at the UN Rio earth summit, it was defined as the use of land resources for the provision of goods or services to meet the dynamic human needs at the same time ensuring the environmental roles of these resources are maintained. In 2005 Terr AFRICA defined it as the "adoption of land use systems through practices of appropriate management that enable land users to maximize social and economic benefits from land while concurrently enhancing land ecosystem functions." SLM deals with water, soil, emission of greenhouse gases (GHG) and vegetation degradation which result to a change in climatic patterns. SLM practices also help to conserve natural resources and at the same time reduce carbon emissions. SLM protects and enhances multiple resources provided to us by land which can be divided into four categories:

- **Continued Provision of Essential Services:** Land provides food, fuel, water, and fodder for animals SLM assists in increasing food security, provide a supply of energy, provide clean water to people and finally support livelihoods, for example, for farmers, miners, construction companies, artists among other professionals
- **Regulation of Services:** These services include regulating water quantity and quality. Regulation of climate and disease control-SLM assists in sequester and store of carbon, mitigation of damages that result from extreme weather and natural hazards, improving water quality and availability and finally pest and diseases
- **Cultural Services:** SLM helps to protect cultural heritage and keep alive cultural and natural landscapes, conservation of indigenous knowledge and cultural practice, and finally the enhancement of aesthetic experiences and create a space for recreation.
- The final category is on support services which includes supporting soil and vegetation cover, water recycling, and nutrient support. SLM will help to resolve the problem of soil degradation and improve soil quality and function, it will also enable nutrient cycling, enhance primary production, and finally provide a home for different species and increase biodiversity.

According to FAO SLM is based on four principles also which include integrated use of natural resources, participatory approaches, policy, and institutional support and finally multiple stakeholder involvement. As a well and widely known organization, it has a mandate and provides support to member countries in the development and implementation of SLMs. It also implements and executes a wide range of programs and projects in relation to SLMs.

2.3.3. Policy Issues in Relation to Sustainable Land Management (SLM)

Sustainable agricultural practices have the ability to reverse the effects of land degradation when applied and as a result they can lead to improved livelihoods, solve the problem of hunger and restoration of natural ecosystems, and finally SLM can contribute to the mitigation of the climate change issue (Letcher, 2007).

Recommendations for policy when implementing SLM practices include the investment in these practices to help increase income, improve food security, and reduce poverty and also reduce global hunger and malnutrition. They also conserve heritage to some extent. Farmers should be educated on how application of SLMs are profitable and support given to smallholder farmers to use appropriate technologies and farming methods. Creation of an enabling environment is necessary for SLM to work.

Another key area is training. It can start by assessing knowledge of farmers on SLM practices and knowledge made available to all stakeholders but especially farmers, policymakers in consultation with researchers and technical specialists. The use of this knowledge can be documented, assessed, and the resultant practices evaluated, and impact measured (Letcher, 2007).

Research can also be another area, although many SLM practices have been documented and their effect determined, there is still the need to carry out more research to clarify the impact of these SLMs in different setups and how to optimize and adapt them. New technologies can also be developed to mitigate these effects.

Awareness creation and capacity building-most technical persons are well informed and have the necessary knowledge on SLMs it is important to pass knowledge to other stakeholders including policymakers who will create legislations on SLM practices and the farmers both small and large scale farmers who will apply these practices.

Provision of agricultural extension services. Most extension services are geared towards increasing agricultural yields which are short term and have been given more priority. Some issues that are sometimes neglected have to do with sustainable use of land resources, preservation of these resources and also ecosystem services strengthening and there is need to provide these services in the future.

Finally, another area is planning and participation. Community involvement and participation is important for the success of SLM. The active participants include extension officers, farmers, researchers, and policymakers. These individuals are involved in planning, development, section, and implementation of the SLMs. For its success, it requires cooperation between community members, provision of appropriate information and knowledge, the exchange of experiences by multiple stakeholders plays a key role (Letcher, 2007).

The activities for SLM can also be planned for but this is the responsibility of policymakers as it is beyond the capacity of just single landowners and users. This can be marked by first mapping areas where land is degraded or has potential to be degraded from this planning for resources both technical and monetary resources can be considered. This can be helpful in visualizing the extent and of intended success and effectiveness of SLM practices. SLMs should be a concern to every individual and if implemented can have several positive benefits.

2.4. DEFORESTATION

Deforestation is the cutting down of forest cover to alter its use by creating space especially for other human activities. According to international forestry research, forests can function to protect people against poverty and also provide at least a quarter household incomes for people inhabiting areas near forests despite these statistics 18 million acres of forest cover is lost to deforestation each year.

2.4.1. Causes of Deforestation

These human activities include conversion of the cleared land to a living space or farmland or even urban use this includes construction, mining, agriculture, and even forest fires. The earth surface covered by forests is approximately 31% and deforestation is carried out mostly in tropical rainforest compared to any other. Deforestation can be as a result of human activities or natural occurrences such as disasters or pest invasion (Figure 2.3).

Figure 2.3: An example of deforestation.

Source: https://www.mnn.com/earth-matters/wilderness-resources/photos/7-striking-examples-of-deforestation-from-nasa/a-planet-in.

2.4.2. Consequences of Deforestation

Environmental effects of deforestation without adequate reforestation have led to loss of biodiversity of flora and fauna, damage to the habitat of wild animals and indigenous communities, climate change and aridity. Deforestation also affects the hydrological cycle since trees are known to extract groundwater making it available which is affected when trees are cut down. One of the most talked-about impacts is global warming which is a result of decreased uptake of circulating carbon and finally soil erosion which leads to the creation of wastelands. Other effects include health effects such as crop up zoonotic diseases, long-term economic losses especially in the cases of overexploitation where the resultant effect is drought or even famine (Letcher, 2007) (Figure 2.4).

Figure 2.4: Effects of deforestation.

Source: https://upscbuddy.com/deforestation-effects/.

There are different NGOs that work towards combating deforestation throughout the world and they also share a similar vision e.g., Greenpeace in the USA, ARAHA in Africa, APECO in Peru, SEPADO in Somalia, Conservation International, Amazon Watch, Rainforest Alliance, Olam international, Mighty Earth, Union of Concerned Scientists, United Nations among others (Harris, 2005).

2.4.3. Policy Issues and Opportunities for Combating Deforestation

Some of the measures policy issues in combating deforestation include reforestation and afforestation which in replanting trees where they were cut down to help restore forests that had been degraded and planting new trees respectively this can be done at both individual level and national level. Different countries can create awareness through the media, within local communities and other social gatherings about deforestation and capacity building to ensure the implementation of tree planting activities. Use of alternative fuel sources to wood or the efficient use of trees as source of fuel.

Respecting indigenous people and their rights, they should be equal rights to other people to help them preserve their culture and avoid the displacement from their homes, e.g., the Cree nation residing in the boreal forest of Quebec. With the ongoing encroachment of forest, local governments need to ensure that their indigenous community livelihoods are preserved, and their heritage remains untouched for their future generations (Harris, 2005).

Governments can also encourage the use of recycled products and in return, this will lower the demand for timber reducing the rate of deforestation at the same time this will reduce land pollution which is done in waste management. People can also be encouraged to buy sustainable wood products.

Fighting illegal logging of trees by encouraging people to report illegal logging and also imposing harsh penalties on illegal tree loggers, people can also employ the use of technology to achieve this in order to protect our forests. Fighting corrupt leaders who are involved in allowing illegal logging to continue. People and the government should not provide support to such leaders

Practicing agroforestry which is a land management system that incorporates planting or growing trees together with plant. This practice has been successful in improving biodiversity, improving soil fertility, and finally preventing soil erosion (Harris, 2005).

We can also provide support to organizations in the fight against deforestation at both national and international level. This can be done by applying sustainable forestry practices, donating to their causes, and engaging in their volunteer programs.

Reducing meat consumption can also be another strategy even though it is impractical with the current food transition towards consumption of more meat products. Since livestock occupy large pieces of land and forest land converted to pasture, people can be encouraged to eat less meat lowering global demand for meat products which will consequently reduce the need to cut down forests to create land for livestock farming.

2.5. DESERTIFICATION AND DROUGHT

Desertification is the process by which fertile arable land becomes a desert while drought is a period in which insufficient rain or precipitation leads to reduced soil moisture leading to crop damage and water shortage. It is an issue that is not limited to one region but it affects different areas globally. Arid, semi-arid, and dry sub-humid areas commonly experience desertification as a form of land degradation. In the southern region of the African desertification can be observed and has 25% loss of soil fertility and approximately a billion tons of soil is lost because of soil erosion. The world day for combating desertification is on the 17th of June and was introduced in the year 1994 to create awareness on desertification and how to combat it by the United Nations Convention to combat desertification (UNCCD) they also brought up ways op desertification prevention and a means of revival from drought. This information is transferred to people with the use of posters, reading material to the public, in school, and in other social gatherings. Their slogan is "let's build the future together" thus the creation of a binding agreement to partner in the achievement of SLM. A more recent agreement between member nations was to combat desertification and the restoration of degraded land. According to UN statistics, arable land use by humans for activities such as agriculture are reduced by 23 hectares per minute driving people in millions into poverty (Harris, 2005) (Figure 2.5).

Figure 2.5: Land degradation from anthropogenic and natural causes.

Source: https://en.wikipedia.org/wiki/Land_degradation.

2.5.1. Causes of Desertification and Drought

Desertification occurs because of climatic variation and activities by humans such as deforestation, urbanization, stripping of land resources considered as unsustainable and are merely not the spread of present deserts. Climatic changes significantly impact drylands because of their vulnerability to desertification due to their low-levels of biological activity or organic matter which increases soil fertility and they are also susceptible to soil erosion and water availability with decreased crop and vegetation cover. Disasters such as a flood will lead to soil erosion reducing soil fertility and reduced precipitation makes land unsuitable from farming.

The human activities that cause desertification include deforestation, over-cultivation, overgrazing, overexploitation of natural resources, and salinization on irrigated cropland. Statistically speaking 35% is attributed to overgrazing, 30% to tree cutting, 7% wood exploitation, 1% industrial activities, and 28% wood exploitation. Intensive cultivation of land induces human-based desertification, this can also be accelerated by a broader problem which is the pressure caused by the growing population and the need to acquire more resources for day to day uses. There is also the migration to drylands which leads to degradation of land. Land can also be degraded if it is left bare and not properly managed. A successful example is Machakos in Kenya where people have managed to regenerate land over the years despite this having a region with land found to be unsuitable for agricultural activities and despite living in adverse weather and climatic conditions they have managed to make economic gains using their natural resources without necessarily over-exploiting them.

2.5.2. Consequences of Desertification and Drought

The negative result is these people then employ intensive farming techniques further degrading land causing desertification and drought. This poses a threat to food security, climate, and the environment. Climate change consequently leads to drought and drought leads to hunger and malnutrition this is therefore a constraint to the achievement of sustainable development goals.

Consequences of desertification also include increased number of unintended urban areas as a result of migration upon land degradation thus leading to overpopulation and problems related to urbanization. They also lead to land conflicts, floods as there is no vegetation to stop water gathering in one place. Agricultural productivity is also reduced and finally there is poor quality of water in the event a piece of land becomes a desert. Increased salinity of soils. Another consequence is reduced flora and fauna species, reduced productivity of land especially dry lands. This loss in the fertility of soil has socio-economic impacts which affect food security and peoples' livelihoods especially those who are dependent on farming as a source of income and this can in turn have other effects such as poverty, migration, and displacement of people that leads to social breakdown resulting in political instability. For example, in 1991 production losses amounted to 42 billion dollars as statistics made by the United Nations Environmental

Program (UNEP). Resilience to climatic disturbances such as drought is reduced (Harris, 2005).

2.5.3. Solution and Policy Areas When Combating Drought and Desertification

There are different solutions for desertification which include policies that can be put in place. The agricultural sector can play a major role in this by changing farming methods and technologies; this includes the intensity of farming on certain areas especially those that have the potential of turning to deserts. Policy changes can be made to the use of land for the extraction of resources such as mining but allowing people to still make a living from this land. Regulation can be put in place to prevent harm and over exploitation.

Education and awareness is important especially in developing countries to help people understand how to best use land for farming and other activities. To save and prevent more land from becoming a desert we there is need to employ more sustainable agricultural practices (Harris, 2005).

Application of new and current technologies in cases where it is difficult to prevent desertification. For such cases, it is necessary to carry out research and determine technologies that can solve this problem. Technological advancements can also help researchers to solve the problem of disasters which also cause desertification.

Preventing deforestation is another way to deal with the problem of drought and desertification this is through afforestation, avoid the cutting of trees in non-forest regions, nutrient cycling, encouraging mixed farming and crop rotation, implementation of agroforestry, water harvesting to also prevent erosion, restriction of grazing to grass and shrubs and also the area and finally building wind barriers.

To correct desertification, we can put up rehabilitation efforts by funding such projects and taking time to work towards corrective measures. By doing this we can prevent these deserted areas from becoming more widespread. Finally, we can apply sustainable practices to prevent desertification e.g., the SLM practices to preserve land (Harris, 2005).

2.6. SUSTAINABLE MOUNTAIN DEVELOPMENT ISSUES

Mountains are one of the natural resources available to humans they function to, provide freshwater as they are sources of water bodies such as rivers, along which people settle and carry out their livelihoods. The world faces different problems such as climate change, natural disasters, population pressure, and loss of biodiversity, political instability, water shortage, and desertification. These challenges affect mountain areas and their inhabitants. Mountains are relied on for resources by about 10% of the world's population. A threat to their presence is the climatic change currently occurring; a range of ecosystems thrive within mountains e.g., tropical rain forests, grasslands, hearth, and more vegetation, lakes that rely on rivers as their inlets. Human activities such as mining and agriculture pose a threat to this ecosystem. With the numerous resources offered by mountains, preserving mountains is an opportunity to achieve sustainable development. When addressing global issues sustainable mountain development is a key area but has been previously marginalized in the agenda for development (Letcher, 2007).

In the year 1992 during the Rio conference by the United Nations, the global community recognized the importance of mountains in the discussion on environment and development. Chapter 13 of their documentation explains the role of mountains within our ecosystem, at the same time stating concerns to do with the degeneration of the quality of mountain but since then not many changes have been made to the issue of sustainable mountain development since it has not been prioritized. Chapter 13 is inclusive of two program areas to explain the problem of mountain ecosystems that are fragile. The program areas include:

- The generation and strengthening of knowledge on the ecology and sustainable development of mountain ecosystems.
- Promotion of integrated watershed development and alternative opportunities to livelihood.

This agenda remains a concern to a few individuals and professionals. As much as this agenda has remained isolated, other agendas such as climate change, biological diversity and conventions combating desertification have made significant progress globally. There are constraints to achieving sustainable mountain development despite the recognition of related issues. Vulnerability of mountain communities and ecosystems can be attributed to the water and natural resources growing demand, pressure exerted by the mining and agricultural industry, food insecurity, conflict, out migration, climate change, and expansion on tourism activities (Letcher, 2007).

2.7. SUSTAINABLE AGRICULTURE AND RURAL DEVELOPMENT ISSUES

Agriculture plays an important role globally both economically and as a source of food. The main source of income for people living in rural areas mostly comes from crop and livestock production. Despite farming, being economically significant farmers lack the knowledge to ensure the agricultural sector is organized in its production, processing, and its sustainability. It is an initiative run and led by civil societies and is supported by government and non-governmental organizations convened by the food and agricultural organization. Farmers require knowledge to manage their farms in a sustainable manner that is ecological, economical, and sociable. At the same time despite having been graced with technological advancements in the agricultural sector there is the threat of hunger with the need to double food production as the population grows (Rangarajan, 2009).

In addition to population growth other natural occurring events such as climate change also threaten agricultural production, land, and soil pollution, degradation of naturally occurring resources also affect sustainable agricultural and rural development. Most productive land is already being cultivated and to ensure the produce is adequate even in the future, we need to ensure we maintain these arable lands. Actively involving farmers in these activities is the key to achieving sustainable agriculture. If the right incentives, knowledge, and government support is provided to farmers then significant progress can be made in the management of water and land sustainability, for example, if farmers grow a number of diverse crops then it would prevent failure in production. In both developed and developing countries there is the need to use new approaches in relation to agricultural systems geared towards the increase of food products while equally protecting resources that farmers depend on. The practices include nutrient recycling, stop the over reliance on certain agricultural inputs such as synthetic fertilizers and chemical pesticides, the use of a diverse number of species while farming, proper management of natural resources, practicing crop rotation or agroforestry to improve and maintain soil fertility. The overall objective of different governments or NGOs should be to optimize agricultural practices taking into consideration the best of both the new and old practices to maximize outputs in a sustainable manner within the available limits of resources (Rangarajan, 2009).

2.7.1. Essential Areas for Sustainable Agricultural and Rural Development

The biological aspects are important and include conserving genetic resources is essential to ensure continued and increased supply of food. Appropriate authorities globally can ensure genetic resources are preserved by conserving the different flora and fauna species (Rangarajan, 2009).

Conservation of soil and water which are physical resources to be able to sustain agricultural productivity. To successfully carry out this SLM practices are required to reduce soil erosion which removes the fertile top soil in order to facilitate the maintenance of soils water holding capacity. Another issue that needs to be dealt with is atmospheric pollution which harms crops and forests such as acidic rain and finally agricultural activity and soil health is affected by excessive use of chemical fertilizers and pesticides.

When property rights are well defined, it acts as a strong incentive to landowners in the usage of land in a more sustainable way. Where land distribution is unfair land tenure reforms are necessary or where there is lack of adequate control on land use. Local controls over planning and resource allocation can be encouraged by local controls (Rangarajan, 2009).

Economically, fair prices in the purchase of produce from farmers and good infrastructure in developing countries, this includes the availability of extension services. Farmers will need an incentive to take part in soil conservation and water resources.

2.7.2. Agenda 21

In 1992, agenda plans were adopted by 169 countries that convened in Rio de Janeiro during the UN conference that was about environment and development (UNCED) which was also called the earth summit. Chapter 21 was devoted to sustainable agricultural and rural development. According to their discussion, major changes were required in agricultural, economic, and environmental policy globally for both developed and developing countries, to develop appropriate conditions for sustainable agricultural and rural development (Harris, 2005).

The agenda 21 also encouraged the integration of sustainable development considerations with agriculture in both developed and developing countries.

It also urges to make sound policies that are current and in line with the current world problems. It also prioritizes the maintenance and bettering of agricultural productivity of arable land that is successful to provide support to the growing world population. Land that does not have a successful past can and should be conserved and rehabilitated and finally the avoidance of marginal land.

Agenda provides the 12 policy areas with detailed proposals. They are encouraging people to make land investments and land reforms by provision of ownership, finances, and ways to the market products at prices that are fair. Training people on soil conservation is also of importance and the combination of both the best of current and old farming methods. There should be use of better ways to conserve genetic resources. Inputs include pest management and control, plant nutrition projects and management, rural energy strategies and technology (Harris, 2005).

2.7.3. Sustainable Agriculture and Rural Development Elements

Integration of activities at government level which include policies, development plans, nutrition surveys, instruments, food quality and security, monitoring, data, and early warning systems.

At community level in the rural areas' development of capacity building and local organizations, allowing active participation of people, training on sustainable agriculture and rural development, provision of extension services essential in achieving sustainable agriculture.

Key natural resources such as land with importance are land use planning, land management, soil conservation, and rehabilitation. The other resource is water with key areas as water conservation, irrigation improvements, water database development, water-users, and associations. Biological plant and animal resources areas include conservation of genetic resources, developing animal and plant breeds. In relation to forests and trees, we focus on reducing deforestation, the sustainable management of forests, habitat conservation, and integration of tree farming systems and finally in relation to fishing we focus our efforts on increase of aquaculture production and preventing overexploitation of water resources (Harris, 2005).

2.8. BIOLOGICAL DIVERSITY ISSUES

Biodiversity is the availability of a variety of biological species, genetics, and ecosystems. Conservation refers to preservation of biological species, their habitats, and ecosystems to prevent them from becoming extinct through their interactions with other biotic and non-biotic factors. Objectives of biodiversity conservation include preserving animal and plant species, creating reservoirs for their preservation, to provide immediate benefits to the tourism industry, act as a future insurance policy, provide knowledge to the scientific community on potential use, and ensure sustainability of different life support systems.

2.8.1. Types of Biodiversity Conservation

There are three types of conservation that is in situ and ex situ conservation and agro biodiversity conservation. *In situ* conservation is the conservation of animal and plant species in their own habitats such as establishing parks and sanctuaries, biosphere, and nature reserves and reserved forests while ex situ entails conserving biological species outside their habitat in places it has become extinct it can include re introducing a species. Agro biodiversity is the conservation of both new and old biological species, the traditional and hybrid agricultural species (Singh, 2008).

2.8.2. Convention for Biodiversity Conservation

In the convention to preserve biological diversity at the Rio conference in 1992 member countries identified ways to have a continuous sustainable supply of biological resources without them becoming extinct. The convention members agreed that the member countries would develop national strategies in order for biological resources to be used sustainable at the same time conserving them. They should also work to create protected areas, the control novel species, sustainable biodiversity, and development of support programs in developing countries. Establishment or research and training programs for conservation and sustainable use of diversity, creation of public awareness and giving public education, apply technology and biotechnology to conserve different species, promotion of technical and scientific expertise between parties, provision of financial resources to provide support for activities (Singh D. A., 2008).

2.8.3. Causes of Loss of Biodiversity

The loss of biodiversity is caused by death of the said species or destruction of the species habitat which results in the reduced survival rates of the biological species:

- They include man-made causes such as the destruction of habitats by human activities such as agriculture and construction.
- Pollution as a result of industrial activities which lead to extinction of species on land, water, and air.
- Global warming affects different biological species.
- Poaching and illegal trade of animals threaten a lot of animal species.
- The use of toxic chemicals such as pesticides and overexploiting wildlife resources leads to a significant decline of some species.
- Loss of habitats rich in biodiversity leads to genetic erosion as a result of the disappearance of indigenous practices of conservation.

Other causes of extinction include ecological factors, pathological, and biological factors due to natural causes.

Some of the species that are being conserved include the white rhino, tigers, protection of the crocodile species *Gavialis gangeticus* in Orissa by the UNDP, a project in 1976 to conserve lesser cat species, 1981 project in Manipur to protect the brow-antlered deer, 1991 project to protect the Asiatic elephant, the 1987 project of Kaziranga to preserve the one-horned rhinoceros, the 1981 project of the Himalayas to prevent extinction of the musk deer and finally project Hangul of 1970 to save endangered Kashmir stag.

2.9. BIOTECHNOLOGY ISSUES

Biotechnology can be applied and used to study the environment. At the same time, it can be implied that it attempts to harness biological processes for exploitation and commercial uses. The International Society for Environmental Biotechnology defines environmental biotechnology as "the development, use, and regulation of biological systems for remediation of contaminated environments (land, air, water), and for environment-friendly processes" (Figure 2.6).

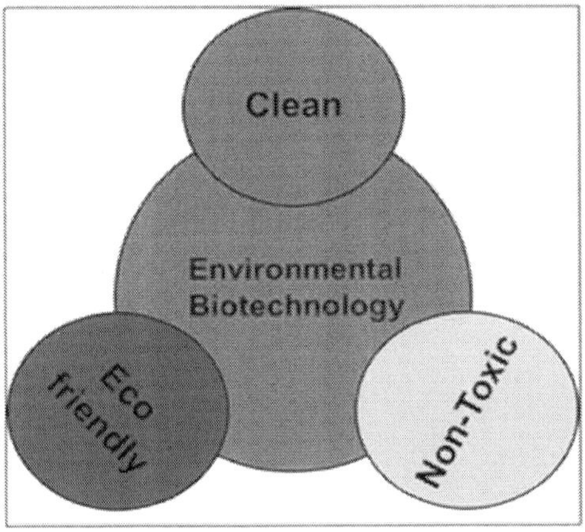

Figure 2.6: Environmental biotechnology safety components.

Source: https://link.springer.com/chapter/10.1007%2F978-981-10-1866-4_1.

We can make use of nature using different life forms such as animals, plants, fungi, and bacteria in the production of renewable energy, and food at the same time making an economic gain in the event where the waste from a process becomes feedstock to another process (Rangarajan, 2009).

The integration of ecological farming systems together with appropriate technology is significant in the achievement of food security, mitigation of climate change and the achievement of sustainable development goals. Environmental biotechnology is of great significance to agro ecology.

Environmental biotechnology has significance to industrial biotechnology, for example, industrially mutated bacteria can be used in other industries such as pharmaceutical industry as they are able to better utilize substrates, microbes can be used to clean oil spills and can be isolated from oil rich environments and yet the argument brought forward is whether this mutated organisms will bring an imbalance in the ecosystem. It is therefore the responsibility of scientists applying the use of biotechnology to the environment to ensure that the relationship between organisms is not altered to maintain the stability of an ecosystem.

The manipulation of genetic materials to create better-fitted species has been ongoing for centuries and has brought with it both positive and negative outcomes to the environment and people's health. It is therefore important to manage these technologies to ensure they cause no more harm (Rangarajan, 2009).

Biotechnology can have foster positive gains in relation to the environment and ecosystems by the development of biological species with new strains that resistant to strains in the environment, reduce costs in agricultural production, higher-quality foods with high nutritive value, waste management, managing forms of pollution and also managing pests and weeds in a benign manner and scientists have been able to achieve some of these milestones

2.9.1. A Precautionary Approach to Biotechnology

A precautionary approach can be used in the management biotechnology in relation to the environment and it stems around four key elements.

The first is to build a scientific base that is credible with a comprehensive way of monitoring the technology and also evaluating its impact. Sound understanding of their effects on human health, productivity, economic, and environmental effects is necessary.

There is also a need to invest in public research and development as an assurance to the public and that the good aspects of biotechnology receive the attention they deserve.

They are the need for reforms especially in institutions with control over biotechnology development to prevent single monopolies from having control and total ownership over technologies especially to certain fields.

Finally, the last is to create a regulatory framework for products from biotechnology. Given the changes made to human health, agriculture, the environment by biotechnology there is a need for governments to evaluate these technologies and regulate access to these technologies by the human population (Rangarajan, 2009).

2.10. OCEANS MANAGEMENT ISSUES

More than 70% of the earth surface is covered by oceans but only a small is marked as protected. It is unfortunate that not most of the oceans have been well managed over the past years. There is a need to carefully plan on ocean protection and management and this should be a joint effort between governments, partners, and other stakeholders globally (Rangarajan, 2009) (Figure 2.7).

Figure 2.7: Marine life; a fish shoal.

Source: http://www.wwf.ca/conservation/oceans/.

The biggest challenge is the harm brought to marine life by human activities which is a threat to their population and marine diversity they include over exploitation through overfishing of marine life and other ocean resources, pollution from both human, agricultural inputs and industrial waste. Other threats to oceans and marine life include global warming, air pollution and invasion by poisonous biological species such as algae which disrupts ecological balance in oceans. Our reliance on marine life for human activities such as recreational activities and tourism, food, and livelihood means we should protect and conserve them to ensure their sustainability.

It also involves the protection of endangered and extinct marine life both plants and animals, for example, Baleen whales were becoming extinct until a ban was placed on their fishing, the green sea turtles and the Kemp Ridley sea turtles have been recognized as two most endangered species. Fish include the pacific bluefin tuna, the American eel, Japanese, and European eel and for marine plants the Johnson see grass (Figure 2.8).

Figure 2.8: A green sea turtle, an endangered species.

Source: https://www.ecowatch.com/endangered-species-act-turtles-marine-mammals-2626362898.html.

2.10.1. How Can We Protect and Conserve Our Oceans?

For the effective conservation of oceans, there is need for the creation of marine protected areas for the management of human activities to enhance sustainability. Organizations such as Conservation International work with governments to conserve oceans that surround us and have adopted the seascape approach where they work with the government and multiple stakeholders to conserve marine ecosystems. It has nine essential elements that is the creation of enabling environments via a legal framework and through adequate institutions, social, and political support, private sector engagement, ecosystem-based management and sustainable financing for effective management, generating outcomes through maintaining and restoration of habitats and ecosystems, species recovery and benefits to human wellbeing (Singh, 2005).

Marine conservation technologies can also be applied in the protection of threatened and endangered species. They increase the survival and health of marine life and still benefits those who depend on them for livelihoods they also include radio frequency identification and the pop-up satellite archival tag it provides scientists with the chance to study marine life in their natural habitat and tracks their movement and the data sent via satellites and is therefore an important tool in studying aquatic animals' behavior. The turtle excluder devices remove threats to turtles especially in the event that they are captured during fishing. There are also laws and treaties that protect oceans and marine life; the most recent one is the International Guidelines for the Management of Deep-sea Fisheries in the High Seas (2009) for the management of deep-sea fishing beyond national jurisdiction.

Other measures include policies by governments to reduce destructive fishing practices an example is trawling, educating fishermen on conservation at the same time helping them to maintain livelihoods, use of measures to prevent accidental capturing of animals other than fish or even fish, minimization of military sonar that kill marine mammals, stop pollution of oceans with industrial and human wastes or treating waste and the use of agricultural inputs that are ecologically friendly and application of biotechnology to clean oil spills hence protecting marine life (Singh, 2005).

2.11. FRESHWATER MANAGEMENT ISSUES

Freshwater sources include rivers, lakes, wetlands, melting glaciers, and groundwater delivered to the surface through springs and wells. These water resources are significant to human survival and livelihood. They provide water for drinking, use in industries, household functions; recreational activities among others. They are also a habitat for animal and plant life which are important as a source of food, tourism activities and consequently livelihood. Human exploitation and other negative environmental impacts such as climate change have resulted in a significant reduction of biological organisms over the years. An estimated 80% of untreated wastewater globally is directed back to water bodies. This in turn affects the health of humans, the ecosystem and the environment as this is the same water that will still be directed for human use. As the population grows, there is also the need for more water which also puts a strain on freshwater resources to meet daily human needs, maintain the aquatic reserve, as well as meet the demands of industries (Harris, 2005) (Figures 2.9 and 2.10).

Figure 2.9: Lake Powel in Utah is a freshwater source.

Source: https://www.nationalgeographic.org/article/freshwater-resources/.

Figure 2.10: A California water source contaminated with sewage.

Source: https://d.newsweek.com/en/full/1531123/sewage-pipe.jpg.

Creation of national and global policies is important in the protection of freshwater sources. There is the demand for integration of planning and management for freshwater resources with the problem of scarcity, gradual degradation, pollution, and a multi-sectoral approach used as different social-economic sectors rely on these freshwater sources for their development and functions. Water conservation and waste minimization measures can be used to support water sources at both household and macro-level (Harris, 2005).

Some program areas have been proposed in the management of freshwater sources by the Socio-Economic Data and Applications Center (SEDAC) of the U.S.

The first program area is the integration of water resource development and management and the main objective of this is to ensure that countries are able to satisfy their freshwater needs in a sustainable way.

The second program area is the assessment of water resources which aims at identifying water sources and their potential, determining how dependable they are, their quality, and which human activities affect them. This information can be used to assess potential conflicts between the supply of freshwater and demand and also provide a database that is scientific for rational utilization of water resources. The third program area is the protection of water resources, quantity, quality, and ecosystems. The fourth program is water and sustainability in urban development, the fifth program is supply of drinking water and sanitation, the sixth, water in both rural development and food production and finally water resources and impacts of climate change on them.

The means of implementation of these programs can be through capacity building, science, and technology, development of human resources and finally financing and cost evaluation.

2.12. TOXIC CHEMICALS ISSUES

Toxic chemicals are used in a variety of work settings. Toxic substances and chemicals may bring to people and the environment in the event of exposure at household and industrial level. Common toxic chemicals include formaldehyde, mercury, lead, asbestos, hazard air pollutants, pesticides, per- and polyfluoroalkyl substances (PFAS), and Polychlorinated Biphenyls(PCBs). Safety measures should be applied when using and handling such chemicals (Singh, 2005) (Figure 2.11).

Figure 2.11: Examples of toxic chemicals in our homes.

Source: https://www.euffslemani.com/examples-of-toxic-substances-in-home.html.

People get exposed to toxic chemicals through drinking water or via the skin, through the air which contains hazardous chemical contaminants, through food, soil, and the effect on health is dependent on the chemical type, the dosage, duration of exposure, and the frequency of exposure. Adverse effects include damage to body organs especially the liver and kidneys some of the substances are also carcinogenic or teratogenic.

Government agencies such as the United States Environmental Protection Agency (US EPA) study and manage extensive databases related to the harm caused to humans with measured dosage of these toxic chemicals. Some regulations include the Toxic Substances Control Program. The US Occupational Safety and Health Administration (US OSHA 2012) also regulates worker exposure to chemicals to prevent adverse health effects (Singh, 2005). The US Congress instituted OSHA under the Occupational Safety and Health Act that President Richard M. Nixon enacted into law on Dec 29th 1970.

To prevent chemical-related hazards manufacturers especially in relation to chemicals are required to determine and evaluate hazards of chemicals they manufacture or import and are required to create label sheets that convey hazard details and how to handle chemicals to prevent contamination and other adverse effects (Figure 2.12).

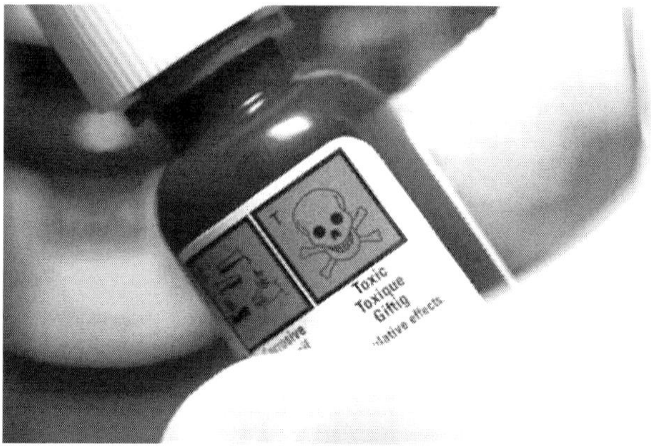

Figure 2.12: A label on a chemical conveying its toxic nature and its effects.

Source: https://www.euffslemani.com/examples-of-toxic-substances-in-home.html.

There are also permissible amounts of chemicals allowed since some of them cannot completely be eliminated including airborne concentrations. The shortcoming of this is that not all chemicals have permissible limits indicated and therefore there is need for more research.

Other alternatives in the use of chemicals can be provided to avoid the use of the toxic chemicals, rotation of job assignments to reduce the frequency of exposure to toxic chemicals. Provision of protective personal equipment is necessary to reduce the risk of exposure and provision of proper ventilation especially for airborne chemical hazards (Uberoi, 2004).

Safe storage of chemicals that are potentially hazardous is of importance and ensuring these containers are labeled appropriately especially at household level. Garments that have come into contact with hazardous chemicals can be disposed of or cleaned separately and finally complete avoidance of the use of toxic chemicals. It is necessary for governments to regulate the use of toxic chemicals available in markets and act to ensure people are not exposed to hazardous chemicals in their workplace, from food, from environmental resources such as water or air etc (Uberoi, 2004) (Figure 2.13).

Figure 2.13: Symbols for toxic chemicals.

Source: https://www.euffslemani.com/examples-of-toxic-substances-in-home.html.

2.13. HAZARDOUS WASTES

Hazardous wastes pose a threat to human health and the environment. They are flammable, have an oxidative effect, explosive, cause infections or cause corrosion. Management of hazardous wastes is purposeful to convert them to wastes that bring no hazard and also stabilize them. Stabilization is done when we handle toxic heavy materials. They can take different forms; they may be solid, liquid, sludges or in gaseous form. They are resultant of the production of chemicals or other industrial activities, medical, and pharmaceutical use and household use. They may cause harm in their storage, transportation treatment or disposal and if not well managed they find a way into water sources and the environment. Governments have created regulations to prevent human health and environmental harm as a result of the poor management of hazardous wastes. These wastes can be toxic even in small amounts; they can be carcinogenic, mutagenic, and cause other adverse health effects (Uberoi, 2004).

There are toxic waste regulations developed by the EPA in the US such as the Resource Conservation and Recovery Act which regulates how waste is handled and disposed. Despite these regulations, there is still the problem of waste disposal in economically disabled areas. The EPA has regulation on biosolids, sewer sludge, and other industrial wastes and with permissible allowance to some wastes such as heavy metals and this has brought up a debate among environmentalists who disagree with this practice with long-term effects in mind. There are systematic regulations on the possession, transportation, handling, and disposal of hazardous waste. Landfills are the most unconventional method of waste disposal and is a practice still seen in developing countries. Preventive and rehabilitative measures are applied in toxic waste management (Figure 2.14).

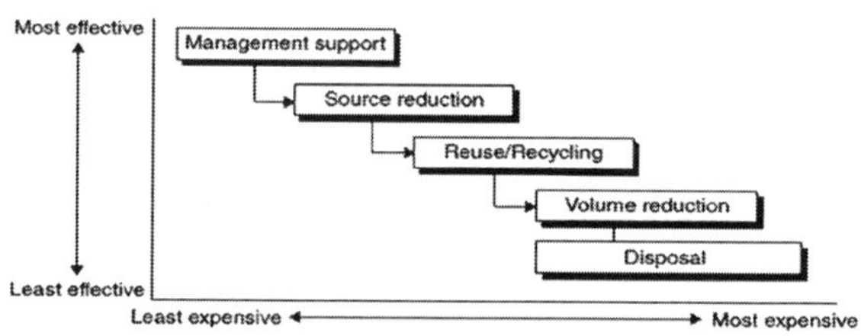

Figure 2.14: Hazardous waste management options.

Source: https://www.intechopen.com/books/management-of-hazardous-wastes/introductory-chapter-introduction-to-hazardous-waste-management.

In the management of toxic waste, the wastes are first analyzed to determine their level of toxicity which will determine how best to transport and dispose these wastes. In the process, some objects can also be recovered when sorting waste and recycled for use in the industrial plant and the cycle continues. Practices in toxic waste management include adoption of strategies such as physicochemical treatment, ensuring regulatory bodies are fulfilling their duties, raising awareness on hazardous waste management, prevention the use of cleaner technologies for toxic waste management, compliance to regulation, creation of new policies when it comes to the management of toxic wastes (Harris, 2005).

In the management of toxic wastes, cleaning of hazardous wastes is an option to be considered; it can begin by planning and financing the activities especially for areas with abandoned and uncontrolled hazard wastes sites. It can begin with investigating the level of toxicity of these wastes and based on the type of contaminants the cost of cleanup is determined. Once cleanup is done, the land can be used if there are no more toxic wastes on site. Phytoremediation is another rehabilitative measure

Despite the regulation put in place in many countries to manage this toxic wastes there is a lack of systematic approach to manage the hazards, they have ineffectively managed to collect and manage the wastes, their regulatory frameworks have failed to address waste treatment and disposal with adequacy and a problem in their implementation. It is necessary to create public awareness in this countries, capacity building, and human resource development for the recycling of wastes that are hazardous, there is also a need to monitor and evaluate programs put in place, provision of funding for such activities and the development of a waste management system for toxic industrial outputs (Harris, 2005).

2.14. SOLID WASTES AND SEWAGE ISSUES

Waste management activities control wastes from their creation to their disposal and this is inclusive of their collection, treatment, transport, and disposal. Solid waste and sewage originate from industries, the environment and from biological species and from households. Wastes can be a threat to human health and to the environment. Waste management is beneficial to humans and the environment and has an aesthetic value. Waste management procedures differ in countries, rural vs. urban areas and also in both residential areas and industries (Figure 2.15).

Figure 2.15: A municipal waste landfill.

Source: http://efc.web.unc.edu/2013/11/05/solid-waste-finance/.

Principles of waste management include efficiency of the use of resources, a products life cycle and the hierarchy of waste management. The rules of waste management are recycling, reusing, and reduction of waste. Waste collection services are mostly provided by local governments or authorities or companies who are private and this is done regularly. Some of the developing countries do not have a well-organized garbage or waste collection system. In some cases, the solid waste at the curb side is separated into recyclable and unusable waste (Figure 2.16).

Figure 2.16: Separation of waste at source.

Source: http://ecosoftt.org/water-wastewater/solid-waste-management/.

Waste can be disposed of in landfills; incinerated, recycled, or reused sewage sludge is produced during treatment of water waste. Common ways of disposing sewage include sanitary landfills, composting, recycling, reusing, and incineration. Another significant method of waste management is preventing creation of waste by reducing it or avoiding it. This can be achieved by buying used products, repairing objects that are broken, avoiding the use of disposable products, creation of products with less material, and clearing remains from containers such as foods.

There are treaties and regulations created for the movement of waste across nations but countries have individual laws to manage their own waste. Key activities to be covered in solid waste management include landfills that are sanitary, investigation of disposal sites, thermal, chemical or mechanical treatment can be considered and an environmental impact assessment (EIA) done (Kulkarni and Ramachandra, 2006).

2.15. RADIOACTIVE WASTES ISSUES

Nuclear industries, the military and medical activities produce radioactive waste. There stages in radioactive waste management which include the planning process, treatment, packaging, storage, and disposal. Nuclear power reactors mostly produce radioactive waste. There are different types of radioactive wastes such low and very low waste, intermediate, and high-level waste. Very low-level waste comes from destroyed material from plasters, metal, bricks, pipes, valves, and activated when these products are dismantled. They are hardly harmful to humans and the environment. These products can either be recycled or used as byproducts. Low-level wastes come from generation by electricity, disease treatment and medical research, smoke detectors, biotechnology, pest control, and the preservation of food. Possess short-lived radioactivity and can be buried in shallow land (Singh, 2005) (Figure 2.17).

Figure 2.17: A worker checks the radioactivity of drums containing nuclear waste at Yonggwang Nuclear Power Plant in Yonggwang, South Korea.

Source: http://www.nbcnews.com/id/13803439/ns/world_news-world_environment/t/asia-has-appetite-nuclear-energy/.

Intermediate level waste may contain large amounts of radioactivity and shielding can be necessary, may come from resins, and fuel metal cladding and chemical sludges. Low-level liquid waste can be discharged from processing plants. High-level wastes derive from uranium fuel burned in nuclear reactors. They are also produced in fuel reprocessing. The fuel is stored in steel-lined containers, water-filled in concrete pools, and tight air containers. Geological disposal is undertaken at 200 meters below ground surface to protect humans and other biotas. Another option for the management of nuclear waste is recycling of the waste through the removal of usable radioactive substances such as uranium and plutonium for energy production.

Reprocessing can be done in a fast reactor to remove uranium, plutonium, and minor actinides. It takes a period of 300 years and is referred to as advanced electrometallurgical reprocessing. There are three kinds of metallurgical treatment at metal smelters and refineries:1, Pyrometallugy using heat to initiate separation of the metals from their mineral concentrate (e.g. copper smelting to produce blister copper, lead smelting).2. Electrometallurgy using electric current to separate the metals (e.g. alumina smelting to produce aluminium).3. Hydrometallurgy using aqueous solutions that dissolve the metal, with sometimes also electrolytic cells to separate them (e.g. zinc production, copper refining). (Singh, 2005).

Some of the proposed methods include chirped pulse amplification which reduces the half-life of radioactive materials from thousands of years to just a few minutes. Other long-term management of radioactive waste includes deep borehole disposal, horizontal drill hole disposal, transmutations space disposal, reusing, and ocean floor disposal (Figure 2.18).

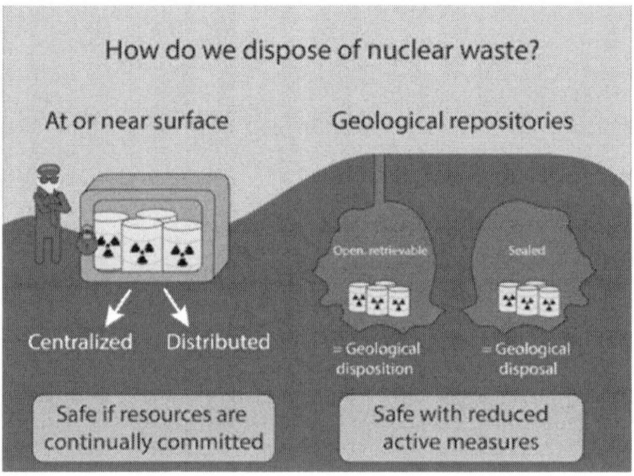

Figure 2.18: Disposal options for nuclear waste.

Source: http://sitn.hms.harvard.edu/flash/2018/looking-trash-can-nuclear-waste-management-united-states/.

3

An Introduction to Materials Management

CONTENTS

3.1. INTRODUCTION

Materials management aims at ensuring the use and reuse of materials in a productive manner over their whole life cycles, with the use of a systematic approach. It helps in representing a change in the manner in which the society thinks about the utilization of natural resources as well as the protection of the environment. By looking at the whole life cycle of products, it is possible to find new opportunities that can help in the reduction of environmental impact, reduction of costs, and conservation of resources.

During the last century, there has been a rapid increase in the global consumption of materials. Research has shown that for the use of raw materials, there has been an increase of 0.4% for every 1% rise in gross domestic product. This increase in consumption has led to a negative effect on the environment, including desertification, overly stressed fisheries, biodiversity loss, and habitat destruction. It has also been noted that an estimated 42% of greenhouse gas emissions in the United States are also associated with materials management. Globally, if we fail to discover ways that are much more sustainable and productive when it comes to the extraction, use, and management of materials, in addition to changing the relationship between the growth and consumption of materials, there will be negative implications for our society and economy (Chatterjee, 2004) (Figure 3.1).

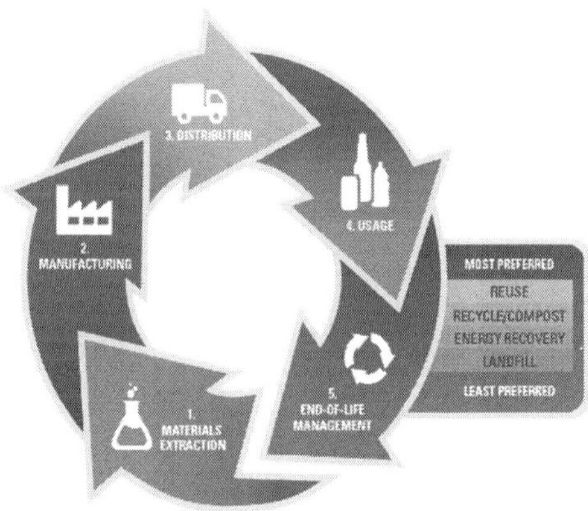

Figure 3.1: Four main stages in a material's life cycle: The extraction of materials, manufacturing, distribution, usage, and end-of-life follow-up.

Source: https://www.epa.gov/smm/sustainable-materials-management-basics.

Sustainable materials management helps in fully representing the framework that should be used to ensure that products and materials are managed in a sustainable manner all through their whole life cycle, which is inclusive of extracting the resource, designing, and manufacturing it, the productivity of the resource, how it is consumed, and end-of-life management.

Firstly the concept of a "throw away society" deals with the large amount of consumerism synonymous with practices and observations of consumer culture in the United States. Secondly the cradle-to-grave principle speaks about the consideration of the life-cycle of a material (usually evaluated through a life-cycle assessment) from a toxicity and environmental impact perspective. The purpose of a cradle-to-grave approach is to challenge the consumerism of a throw-away society.

Sustainable materials management helps in representing a shift when it comes to how we can use and value materials while still focusing on the impact they have on the environment, and how we can protect the environment by ensuring that we pay attention to the whole lifecycle of the product. There are many governments all around the world, including the US environmental protection agency, that have adopted the sustainable materials management approach when it comes to the management of materials (Chatterjee, 2004).

There are many concepts and programs all over the world that are currently being used, and are either being overlapped or supplemented by the sustainable materials management approach. These are inclusive of OECD's Sustainable Materials Management framework, UNEP's Sustainable Production and Consumption and Sustainable Resource Management programs, the G8's 3R's (Reduce, Reuse, Recycle) program, the US EPA's Design for the Environment Program, green procurement, lean manufacturing, sustainable supply-chain management, eco-labeling, green chemistry, and zero waste.

We need to understand the differences between sustainable materials management and waste management. Some of these are:

- Sustainable materials management aims at optimally using all resources all through their lifecycle, while waste management only aims at the management and reduction of waste and waste pollutants at the final stage of the life cycle (Figure 3.2).

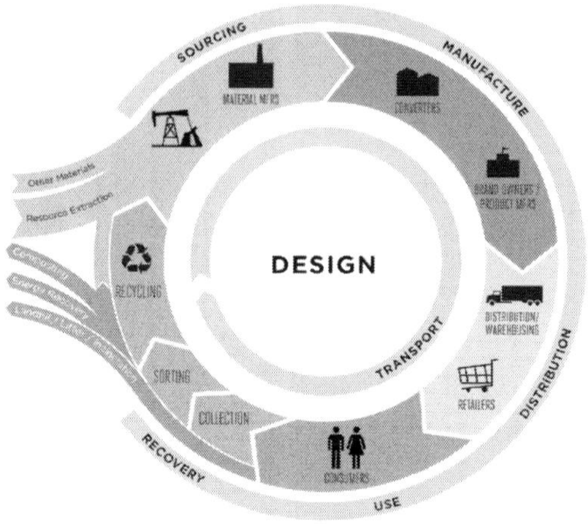

Figure 3.2: Sustainable materials management-another critical step towards waste minimization.

Source: https://nerc.org/news-and-updates/blog/nerc-blog/2015/03/10/sustainable-materials-management

Sustainable materials management pays attention to the whole lifecycle of the product and the material, while waste management only deals with the waste product's end-of-life management.

- Sustainable materials management deals with output and input from/to the environment, as well as any impacts that are in association when it comes to water and air, and it has no geographical constraints. Waste management on the other hand only deals with the outputs from waste to the environment, and the places in which waste management occurs.
- The main goal of sustainable materials management is long-term sustainability, while waste management aims at the management of the impacts on the environment that are due to waste products.
- Sustainable materials management puts into consideration all consumers and industries that are in association with the lifecycle of a product or a material as parties that should be held responsible when it comes to the management of materials and products, while waste management only considers those who generate the waste as parties that should be held responsible.

3.2. SUSTAINABLE DEVELOPMENT

Sustainable development refers to the main principle that should be used to meet the human development goals while still being able to sustain the ability that natural systems have when it comes to providing ecosystem services and natural resources, all which the society and economy can depend on. The aim is to reach a position where the resources and living conditions of the society are used in a manner that still ensures that the human needs are met without undermining the stability and integrity of the natural system. We can define sustainable development of development that is able to meet the present needs without affecting the ability of the future generations to have their needs met (Figure 3.3).

Figure 3.3: Above is a pictorial illustration of the sustainable development goals. All of these are important in achieving sustainable development. It all involves the promotion of economic development without depleting the essential natural resources.

Source: https://www.un.org/sustainabledevelopment/blog/2015/12/sustainable-development-goals-kick-off-with-start-of-new-year/.

Despite the fact that we derived the modern concept of sustainable development from the 1987 Brundtland report, the concept has also been advanced from previous ideas about the 20th century environmental concerns and sustainable forests management. As we have been developing the concept, there has been a shift that has led to the gravitation towards the development of the economy, protection of the environment for future generations, and social development. There have been suggestions that we should use the term sustainability to refer to target goal that we have set for humanity in achieving a human-ecosystem equilibrium, whole the term sustainable development should be used to refer to the temporal processes and holistic approach that will help us achieve the end point of sustainability (Boamah, 2017).

Both the obligations that have been set to preserve ecosystems and natural resources and the ambitious economic development that is found in most modern economies have been seen to have a conflicting nature and they are as of now endeavoring to reconcile them. There is an unbalanced opinion that tries to prove that as a remedy to economic development, rather than holding the commitments that have been set against climate change and other sustainability measures, leveraging, and turning them into market opportunities may have a greater benefit. When such organized practices and principles are used to bring about economic development in an economy, this is referred to as managed sustainable development.

There has been a lot of criticism on the concept of sustainable development, including the question of exactly what is meant to be sustained in sustainable development. There are some arguments that have stated that it is not possible to sustainably use a nonrenewable resource, because any positive exploitation rates will end up leading to the exhaustion of the stock on earth, and this means that the argument therefore renders the whole industrial revolution unsustainable. Another argument states that the meaning of the concept has been stretched to economic development from conservation management in an opportunistic way, and the report did not promote anything but a business as the main strategy for developing the world, with an insubstantial and ambiguous concept attached as a slogan for public relations (Boamah, 2017).

There are three main components of sustainability: the environment, the community, and the economy. These three are interrelated and collectively and individually important in obtaining a community that is sustainable. When you come to balancing the demands that each company has to provide, it has proven to be a key challenge in the creation of global plans. The three components are explained below:

- **Environment:** We are able to achieve environmental sustainability through being stewards of the natural environment and also in ensuring the reduction of the negative effects that human activities have on natural systems and resources that are meant to support the global community (Figure 3.4).

Figure 3.4: When nature is left undisturbed, it is able to take care of itself and survive; however, when we come into the picture and excessively utilize the Earth's natural resources, we damage nature's ability to take care of itself. If we do not use methods to achieve environmental sustainability, we will compromise long-term viability of the natural balance of the ecosystems.

Source: https://careerbright.com/career-jobs/the-role-of-electricians-in-promoting-a-sustainable-environment.

- **Community:** A community that is sustainable is one that is safe, promotes well-being, diverse, healthy, and accessible. Communities are achieved by supporting the arts, historic preservation, and cultural diversity, access to social services, healthy living, and public participation.
- **Economy:** For an economy to be sustainable, it needs to be resilient and strong. A sustainable economy can be achieved by supporting economic development, innovation, green industries, businesses, jobs, and education.

3.2.1. Important Elements in Achieving Sustainability

- **Regulations:** These are normally set in place to provide protection to the environment and even individuals. It doesn't matter what the topic is, whether it's consumption of goods and services, safety, and health in the workplace or home, environmental protection, regulations have proven to be very important at providing positive effects in achieving sustainability. Regulatory processes help in ensuring that federal organizations are able to encourage sustainable practices that are eco-friendly, and also to promote behaviors that are sustainable.

Our collective attention has been directed to environmental systems and also the need for the development and improvement of our ability to use sustainable methods to correct the many different negative environmental impacts due to certain natural and man-made activities such as illegal fishing, poaching of indigenous species, natural disasters like tsunamis, and climate change (DATTA, 2008).

In many different countries, federal energy and environmental regulations have helped in educating the public on the issues related to sustainability and to the creation of normative responses that help in widening and promoting the application of sustainable procedures and practices within the government. For example, there are certain executive orders that agencies are required to follow, and despite the fact that they are able to choose different means to do so, they are able to often discover different innovative means, during the process of compliance and implementation, that help in contributing to the overall policies for sustainability (Figure 3.5).

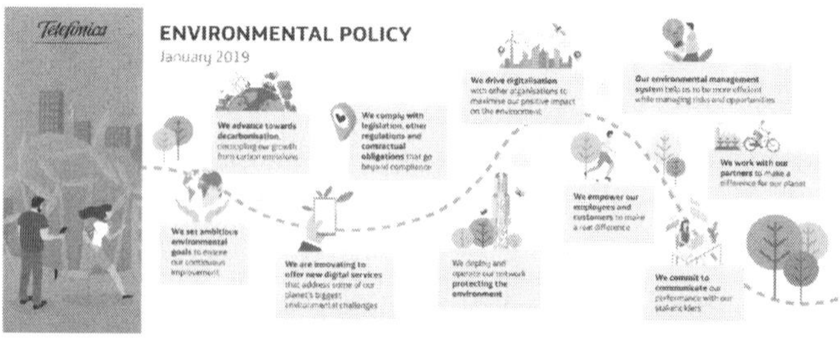

Figure 3.5: An illustration of a corporate plan to develop and implement policies in stakeholder engagement, change management and sound environmental management.

Source: https://www.telefonica.com/web/responsible-business/environment/environmental-management.

Promulgation of eco-friendly regulations helps in the achievement of sustainability goals. As of today, agencies are applying eco-friendly practices across a wide range of regulations. Within the federal community, by the time environmental sustainability was being used as a common term, environmental regulation had already become widespread. Many of the policies, procedures, and processes that have led to actions that have been considered sustainable, have shown a pattern within the government that is similar to that which was used during the development of many different technology-forcing, administrative, and environmental regulations. These patterns have proven to be very useful as referrals in achieving sustainability goals. For example, efforts that have been put in place to prevent pollution with the aim of reducing emissions from industrial sources have ended up creating actions that have proven to be sustainable in preventing other negative environmental effects from occurring in the future.

Many activities and processes that federal governments are expected to undertake have been addressed by environmental regulations thus allowing them to accomplish sustainability goals while still being able to achieve their environmental goals. Procedures and policies are part of regulatory processes all applying sustainable practices to "green" procurement practices, daily mission activities, and industrial processes (DATTA, 2008).

Friendly competition among agencies is encouraged by environmental and sustainability goals, which results in new insights and innovative solutions, leading to a further extension to achieving sustainability. Quantitative environmental measures that have already been established can also help in increasing sustainability practices. For example, having five years of environmental data can help in providing five years of data that is very helpful in publicizing sustainable processes and practices, therefore making it much easier for other agencies to use as a guide, and also in supporting individual plans for sustainability performance that may be needed by the office of management and budget, and by the council on environmental quality (Figure 3.6).

Figure 3.6: The success rate of regulations is much mood than just compliance. Complying with environmental policies and regulations is a building block that is considered very important in environmental sustainability.

Source: https://www.lanl.gov/environment/protection/compliance/index.php.

It is also possible for agencies to require or promote the use of appropriate practices in their various sectors, so as to make sure that they have effective, efficient, and economical sustainment of resources that they commonly utilize, in the advancement of corporate goals. There are many simple actions that are considered as appropriate practices, and these range from the reduction of petroleum use and even replacing the use of petroleum with newer fuel types as a way to lower the consumption of energy while still being able to ensure that sustainability goals are advanced, reducing toxic contaminants during the use of materials, and even being able to recycle electronic equipment through the development of complex systems for solar, wind, and geothermal energy development. Proactive use of environmental best practices, protection of various species, and environmental conservation and preservation all contribute to sustainability goals (DATTA, 2008).

There are also other environmental actions that are categorized as sustainable practices including: preservation of historic resources, protection of certain species, conservation of water and energy, and recycling materials, all playing a huge role in achieving sustainability goals in their implementation and intent.

3.2.2. Consumer Awareness

Although sometimes unknowingly, the driver of production remains to be the consumer. This can be very dangerous considering the fact that many consumers do not know that they have this responsibility. Consumers have continued to make promises in ensuring the improvement of the environment, however, they have developed fatigue in following up on these promises, and therefore leaving nature to take its course. There has been a shift in the focus when it comes to the demand for accountability on the habits of consumption from consumers on this earth that can feel and show the effects of human activity.

Consumption habits have been the main primary driver for climate change. The drive for economic growth globally, increased infrastructure and expansion of cities has ended up changing how consumers view life. Services and products have been made much more accessible and affordable by producers without considering the effect they may bring to the environment due to waste management. Despite a global campaign to reduce the production of plastic, they are changing forms. In almost all public gatherings and boardroom meetings, plastic bottled water has become the most popular drink; however, there are no sustainable recycling systems available for ensuring proper waste management of plastic bottles (Mishra, 2009) (Figure 3.7).

Figure 3.7: Above are the rights that consumers have. Educating the public on these rights can be a step towards building appropriate habits when it comes to environmental sustainability.

Source: https://www.facebook.com/pages/category/Education-Website/Consumer-right-awareness-242872294385755 7/.

Consumers have five main responsibilities. These include: Lawfully purchasing goods and services, speaking out against wrongdoing, using products and services in a proper manner, reading, and following instructions, and staying informed. There are many different countries where consumers pay little attention to the amount of packaging that becomes poorly managed waste.

As consumers, we tend to spend a lot of time in our personal cars in traffic, so as to keep up with the dictated status, and yet if there was efficient management of public transport like in Western countries, we would be able to deal with at least half of the issues in regards to pollution, health problems and time-wasting all which are associated with sitting for long hours in tiny cars and vans. If consumers were wise, they would consider changing to a train coach which is much more comfortable, spacious, and safer, and not only would it use one straight path to a certain destination, but also lower the level of stress that we assume in traffic (DATTA, 2008).

There are eight main consumer rights. These are: The right to a healthy environment, the right to consumer representation, the right to consumer education, the right to redress, the right to choose, the right to information, the right to safety, and the right to satisfaction of basic needs. If these rights were properly exercised, the world would immensely improve.

When it comes to going green, there are many citizens who have no idea what it means, and thus it remains a foreign phenomenon. We should be able to embed lessons on the making of decisions that are environmentally friendly in the teaching foundation in schools, for example, 'reduce, reuse, and recycle.' It is important to ensure that we give school going children early guidance on conscious living, which results in bringing up consumers that are sustainable. We need to use role models that switch off water taps and lights when they are not in use, eat organic foods from farms, and cycle to work. We will also be able to create new opportunities and new jobs if we manage the limited resources that we have globally (DATTA, 2008).

It is important for leaders to include consumer education in the campaign so as to inform their voters on how they can use water and farm sustainably so as to conserve energy and promote environmental sustainability, especially in third world countries, so as to avoid unexplained power shortages and dry taps.

There are many people in the world who are not able to get their basic needs. Those who have the job of representing these people need to come up with solutions that are long-lasting, so as to ensure that everyone's basic needs are accounted for. There is a need for new space and ideas for the newer generations to make explorations of better habits for consumption. It is important for all countries to consider running a sustained campaign on the creation of awareness on the rights and responsibilities that consumers have, and this will help in increasing the level of accountability among regulators, producers, and consumers.

3.2.3. Companies' Solution of End-of-Pipe Problems

When we talk about end-of-pipe solutions, we are referring to an approach for pollution control that helps in cleaning up flows of air or water that are contaminated at the point of environmental entry of effluents. It is possible to make a comparison with other routes, which is often less, costly, of the prevention of contamination in the first place. It is actually much better to deal with the latter strategy of dealing with the contamination before it gets into the pipe. This approach is much more cost-effective and elegant, however, it is much more difficult to regulate or dictate in comparison to the control technologies that are for the end of the pipe (DATTA, 2008).

For this approach, stormwater is a good example. Some research has been done, and stormwater can be managed in two main ways: controlling the water source which involves making sure that stormwater doesn't runoff and also making sure that there are no pollutants that enter the runoff; and treating the water, which involves the elimination of pollutants from the flow of stormwater.

Due to environmentalists, we have been able to find out that there is a different answer to pollution that does not deal with before the pipe or the end of the pipe. It actually does not involve the pipe at any point. Instead, it eliminates the effect of the pipe with just a pen stroke: Donations. There are many companies that donate emission credits for which they get a deduction on tax (DATTA, 2008).

3.2.4. Companies' Green Product Development

Green products are products that are sustainable and have been created specifically for minimizing the negative environmental effects that certain materials cause during the entire life cycle even after they are not eligible for use anymore. There are two main goals of green products. These include: maximizing resource efficiency and reduction of waste. They normally use procedures that are environmentally friendly and also toxic-free ingredients to manufacture the products, all which are certified by organizations that have been recognized, such as the Forest Stewardship Council, Energy Start, etc. (Figure 3.8).

Figure 3.8: Above is a schematic representation of green product development. Protecting our earth, sustainable development, sustainable lifestyle, environmental protection, going green, green revolution, among others has become an important aspect of our daily lives.

Source: https://www.researchgate.net/figure/Fig-No3-Green-Product-Development-Cycle_fig3_282855451.

Green products have certain characteristics. These include:

- When growing them, there is no use of toxic chemicals, and they are also grown in conditions that are hygienic.
- These products are biodegradable in nature, can be reused, and they also can be recycled.
- The products come in a package that is eco-friendly.
- It does not require many resources.
- It is eco-efficient.
- The level of carbon footprint is lowered or nil.
- The level of plastic footprint is lowered or nil.
- Generally, companies that produce green products normally communicate using green marketing to inform of their value proposition to the market.

Green products have many advantages, including:

- Brands that go green are able to have access to new markets: when brands develop green products, it helps with opening doors to many different new markets of consumers who normally only purchase green products while probably still paying more for them.
- Going green provides a competitive advantage: there has been evidence that going green has been able to provide brands with a higher competitive advantage in a market that is trying to reduce the level of climate change.
- Going green helps in improving your public image: there is always an enhancement of the brand image, especially when it does a good thing for the environment and the society.
- Going green gives you loyal customers: there are many individuals that generally prefer to use green products in comparison to the conventional products that are not environmentally friendly, and going green automatically helps in gaining this set of loyal green customers.

Customers can also benefit from going green. Some of the advantages include:

- The products are cost-effective: in comparison to conventional products, green products last longer. In addition, green products normally consume a lower amount of energy and other resources, therefore lowering user bills. For example, using solar electricity by installing solar panels helps in lowering electricity costs.

- They are easy to maintain: when you maintain and operate green products in a responsible manner, it can lead to lower maintenance costs. For example, the use of green buildings means that not only will they be able to market themselves due to the fact that they are environmentally friendly, but also the fact that they have lower operational costs.
- They help in improving health: due to the fact that all green products are created from materials that have no toxic products (components and chemicals), they help in improving mental and physical health. Large windows are normally installed in green buildings allowing for fresh and healthy air, together with abundant natural light. This may help in the reduction of stress and improvement of mental health (Figure 3.9).

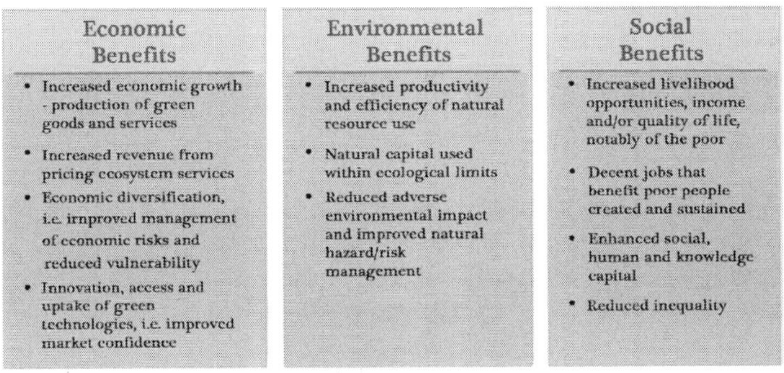

Figure 3.9: These are social, environmental, and economic benefits of going green. Going green not only leads to increased economic growth, but also economic diversification, which involves reducing the level of vulnerability and improving the management of economic risks.

Source: http://ap.fftc.agnet.org/ap_db.php?id=794.

In addition to all these advantages, there are also advantages to the society. These include:

- **Generation of More Jobs:** A study has proven that industries that create renewable energy have helped in increasing job opportunities by 5.3% between the years 2016 and 2017. Predictions have been made showing that with increase in demand of green products, there will come an increase in job opportunities, rising to 16 million by 2030. This means that developing green products not only improves the environment but also improves the economic conditions.

- **Prevents Resource Overuse:** Development of green products helps in encouraging the generation of energy with the use of natural resources and therefore does not threaten fossil fuels and resources through overuse.
- **Protection of the Environment:** Industries normally use materials that are biodegradable and organic to create green products and they are also designed to use the least amount of toxic chemicals and nonrenewable resources for the production of energy. Using these methods, helps in the reduction of greenhouse gas generation such as methane, Ozone, CFCs, etc. Therefore, lowering the level of pollution and also preventing climatic deterioration.

Despite the fact that all these advantages exist, there are also a couple of disadvantages that we have to consider when going green. These include:

- **The Products Are Costly:** For industries to develop green products, they need a lot of investment and innovation. This leads to an increase in the cost of product development, which ends up making the products much more expensive in comparison to the conventional alternatives that are found in the market. The high costs normally lead to lack of customers because it discourages the consumers from buying them.
- **There is a High-level of Ignorance from Consumers:** There are not many people who are fully aware of the advantages and the importance of going green. Majority of the masses are still ignorant.
- **Investment:** New technology has to be developed for the production of green products. This means that a large investment has to be put into research and development. There are not many industries and startups that are able to risk spending that amount.

3.2.5. Principles of Stockholm Declaration on Environment

This is also referred to as the declaration of the United Nations Conference of the Human environment. Its adoption occurred in 1972 on June 16th. In international environmental law, it was the first-ever document to make a formal recognition of the right to a healthy environment. This declaration meant that nations were in agreement when it came to accepting the responsibility that came with causing any negative environmental effects.

Seven proclamations within the document covered the nature of environmental problems globally in addition to stating that they need a high-level of cooperation among the nations to deal with it. In conclusion, the proclamations were able to state that the work put into improving and preserving the environment will be advantageous to the human race and its prosperity. Twenty-six principles were then stated in the declaration so as to help in guiding the nations through the tasks they had to accomplish (Figure 3.10).

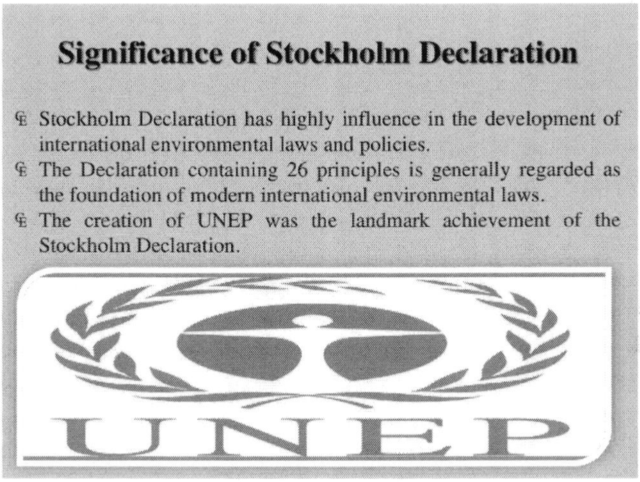

Figure 3.10: The Stockholm declaration has been very important when it comes to developing international policies and laws. The 26 declared principles are considered the foundation of all the international environmental laws we have today.

Source: https://pt.slideshare.net/AbuKhairulBashar/stockholm-declaration/3.

Some of the issues that were discussed at the conference include:

- The cultural and social aspects of environmental issues.
- Measures of pollution control.
- Management and problems of human settlements.
- Development and how it relates to the environment.

- Natural Resource Management.

The 26 principles include:

- **The Right to Protect the Environment:** Human beings have fundamental rights to adequate life conditions, equality, and freedom, in an environment that is able to provide a dignified life and appropriate wellbeing, while still bearing the responsibility of protecting and improving the environment for the generations that will come in the future.
- **Managing Natural Resources:** The earth's natural resources including fauna and flora, land, water, air, and most importantly representative samples of natural ecosystems need to be protected so as to ensure that the future generations also benefit from them through careful and appropriate management and planning.
- **Managing Renewable Resources:** It is important to maintain the earth's capacity to produce important renewable resources, and if possible, this capacity should be restored and improved.
- **Conserving Wildlife:** The heritage of wildlife and its habitat needs to be wisely managed and safeguarded by us human beings, especially due to the fact that they are at great risk of becoming endangered due to the combination of adverse factors.
- **Managing Non-Renewable Resources:** It is important to ensure that we employ non-renewable sources in a manner that can ensure that they are protected from future exhaustion, in addition to making sure that their advantages are enjoyed by all.
- **Controlling Pollution:** So as to ensure that there is no infliction of serious or irreversible damage upon ecosystems, discharging substances that are toxic or other harmful substances, and also releasing heat that the environment is not able to contain was ordered to end.
- **Preventing Sea Pollution:** Nations were asked to ensure that seas are not polluted by substances that can damage amenities, harm both marine life and human health, or cause interference with other important functions of the sea.
- **Social and Economic Development:** So as to ensure favorable working and living conditions, social, and economic development is important. It is also needed for the creation of the appropriate conditions that are required for improving life quality.

- **Natural Disaster and Underdevelopment:** Natural disasters and underdevelopment are able to create environmental deficiencies. These can however, be fixed through the transfer of enough technological and financial assistance.
- **Stable Primary Commodity Prices:** For countries that are developing, it is important to have stable prices and enough earnings for the raw and primary materials so as to be able to efficiently and effectively manage the environment.
- **Environmental Policies:** The present and future developmental potential for countries that are developing should not be negatively affected by environmental policies; however, they should be enhanced. A coordinated and integrated approach should be adopted by states when it comes to development planning.
- **Environment Protection Education:** There should be available resources for the preservation and improvement of the environment. When it comes to environmental matters, education is essential and nations should receive help from mass media to achieve this.
- **Management of Resources in a Rational Manner:** It is important for states to adopt a coordinated and integrated approach to their development planning so as to achieve the rational management of resources.
- **Rational Planning:** This is an important tool that helps in the reconciliation of conflict that may arise between the need to ensure protection and improvement of the environment, and the needs of development.
- **Human Settlements:** Urbanization and human settlements need to be planned in an appropriate manner so as to ensure that states avoid causing any adverse effects on the environment.
- **Human Population:** In regions where the population growth rate is likely to cause a negative effect in the environment, it is important to apply demographic policies that have no prejudice to the basic human rights and they also have to be confirmed by the concerned government when it comes to their validity.
- **Setting up Environmental Pollution Control Agencies at the National Level:** The task of controlling, managing, and planning for environmental resources needs to be entrusted to appropriate national institutions.

- **Using Science and Technology:** When it comes to identifying, avoiding, and controlling environmental risks, it is necessary to apply science and technology. It is also applicable in finding solutions for environmental problems.
- **Providing Education in Matters to do with the Environment:** So as to broaden the number of people who are enlightened on matters to do with environmental sustainability, provision of education in environmental matters is very important. It also helps in ensuring that communities, enterprises, and individuals conduct themselves appropriately when it comes to improving and protecting the environment as a whole.
- **Furthering Scientific Research:** Especially in developing countries, it is important to promote scientific research and development so as to find solutions to environmental problems.
- **The Rights and Responsibilities of Sovereign Nations:** Countries have the sovereign rights when it comes to the exploitation of their own resources in accordance to their own policies to do with the environment, as well as the job of making sure that the activities that they are performing are not damaging both the environment and neighboring countries and states (Figure 3.11).

Figure 3.11: This was the first time that environmental issues were put on the national agenda. The progress for environment and development was started at this conference which happened in 1972 from the 5th to 16th of June. One of the most important things that happened at this conference was the creation of the UN environmental program (UNEP).

Source: https://pt.slideshare.net/AbuKhairulBashar/stockholm-declaration/3.

- **Developing International Law:** In regards to compensation and liability for those who have been affected by pollution and other damages due to the environment, countries were expected to cooperate in the development of newer international laws.
- **Implementing the Agenda in Every Country:** Every country is expected to consider, without prejudice, the systems of values that are essential in each of the countries, as well as the level to which the standards are applicable in the developed countries, and those that are inapplicable in developing countries.
- **International Cooperation:** It is important to handle international matters that have to do with protecting and improving the environment in a cooperative spirit and on equal footing by all countries no matter their size.
- **Dynamic and Coordinated Roles:** Countries are expected to make sure that international organizations play a dynamic, efficient, and coordinated role when it comes to protecting and improving the environment.
- **Nuclear Weapon Ban:** No human being and environment should have to go through the effects of nuclear weapons. All should be spared. This is also inclusive of other methods of mass destruction.

3.3. STAKEHOLDER CONCEPT

Businesses are now taking up strategies that not only cater for the basic requirements of the enterprise and the business operation, but also those that comply with the regulations that have been set for environmental protection and development, and they have also started to recognize that sustainability is paramount in protecting the human existence and natural resources. Sustainability ensures that the future impact of economic development is taken into account and works by incorporating it in a manner that will be advantageous to the environment, the company, and the consumer's quality of life.

When we use the word stakeholder, there are many people involved. These include: employees, shareholders, the community that is interested in the development plans and strategies of the company, and also the customers. All of the stated people are affected in one way or another by the sustainability efforts the company puts in, and in the end, it is a trickling effect, because those efforts end up affecting the global environment and society as a whole (Mishra, 2009).

Due to the fact that businesses are highly linked to developing countries and the global economy, stakeholders are inclusive of citizens that reside in developing countries. Sustainability efforts and multinational business strategies by corporations greatly affect these individuals, because natural resource reserves and water quality is improved by environmentally sound practices.

Multinational corporations now understand that there is a major concern when it comes to wasteful business practices and the environment, thus making sustainability a major trend today. Majority of the consumers are now more mindful of eco-friendly consumption, carbon footprint, emissions, and air quality, thus making corporations much more careful in adopting advertising campaigns that aim at creating assurances to the consumer on their commitment to maintaining sustainable policies. There are some business sectors and employees who have been affected negatively by the demand for eco-friendly products because some of the manufacturing processes that were initially used have become obsolete (Agarwal, 2005).

When it comes to improving the quality of life and adhering to environmental regulations, corporations are increasingly being held accountable to their stakeholders. Most companies in the past only cared about making sure that the shareholders and investors benefit economically, and despite the fact that there are many corporations that still operate as such; stakeholders are admitting that sustainability can be very beneficial without causing severe damage to the environment (Agarwal, 2005).

3.4. ENVIRONMENTAL PROBLEMS

Environmental problems refer to the negative effects that occur on the biophysical environment due to human activity. The act of protecting the natural environment on all levels including governmental, organizational, and individual, is referred to as environmental protection. This is all done with the purpose of benefiting both humans and the environment. Issues to do with the environment are addressed through activism, education, and advocacy in a sector referred to as environmentalism.

The carbon dioxide equivalent of greenhouse gases (GHG) in the atmosphere has already exceeded 400 parts per million (NOAA) (with total "long-term" GHG exceeding 455 parts per million) (Intergovernmental

Panel on Climate Report). The amount of greenhouse gas in the atmosphere is possibly above the threshold that can potentially cause climate change. The UN Office for the Coordination of Humanitarian Affairs (OCHA) has stated "Climate change is not just a distant future threat. It is the main driver behind rising humanitarian needs and we are seeing its impact. The number of people affected and the damages inflicted by extreme weather have been unprecedented." Further, OCHA has stated:

Climate disasters are on the rise. Around 70% of disasters are now climate-related-up from around 50% from two decades ago.

These disasters take a heavier human toll and come with a higher price tag. In the last decade, 2.4 billion people were affected by climate-related disasters, compared to 1.7 billion in the previous decade. The cost of responding to disasters has risen tenfold between 1992 and 2008 (Chatterjee, 2004).

Destructive sudden heavy rains, intense tropical storms, repeated flooding, and droughts are likely to increase, as will the vulnerability of local communities in the absence of strong concerted action. Environment destruction caused by humans is a global, ongoing problem. By the year 2050, the global human population is expected to grow by 2 billion people, thereby reaching a level of 9.6 billion people. The human effects on Earth can be seen in many different ways. The main one is the temperature rise, and according to the report" Our Changing Climate," the global warming that has been going on for the past 50 years is primarily due to human activities. Since 1895, the average temperature in the United States has increased by between 1.3°F and 1.9°F, with most of the increase taking place since around 1970 (Chatterjee, 2004).

3.4.1. Types of Environmental Problems

Major current environmental issues may include climate change, pollution, environmental degradation, and resource depletion. The conservation movement lobbies for protection of endangered species and protection of any ecologically valuable natural areas, genetically modified foods, and global warming:

- **Climate Change:** Global warming is the mainly human-caused rise of the average temperature of the Earth's climate system and has been demonstrated by direct temperature measurements and by measurements of various effects of the warming. It is a major aspect of climate change which, in addition to rising global surface temperatures, also includes its effects, such as changes in precipitation. While there have been prehistoric periods of global warming, observed changes since the mid-20th century have been unprecedented in rate and scale.

The Intergovernmental Panel on Climate Change (IPCC) concluded that, "human influence on climate has been the dominant cause of observed warming since the mid-20th century." These findings have been recognized by the national science academies of major nations and are not disputed by any scientific body of national or international standing. The largest human influence has been the emission of GHG such as carbon dioxide, methane, and nitrous oxide. Fossil fuel burning is the principal source of these gases, with agricultural emissions and deforestation also playing significant roles.

The effects of global warming include rising sea levels, regional changes in precipitation, more frequent extreme weather events such as heatwaves, and expansion of deserts. Surface temperature increases are greatest in the Arctic, which have contributed to the retreat of glaciers, permafrost, and sea ice. Overall, higher temperatures bring more rain and snowfall, but for some regions, droughts and wildfires increase instead. Climate change threatens to diminish crop yields, harming food security, and rising sea levels may flood coastal infrastructure. Environmental impacts include the extinction or relocation of many species as their ecosystems change, most immediately in coral reefs, mountains, and the Arctic. Some impacts, such as loss of snow cover, increased water vapor, and melting permafrost, because feedback affects that further increase the rate of global warming. Ocean acidification caused by increased CO_2 levels is commonly grouped with these effects even though it is not driven by temperature. Ocean acidification is happening because increased carbon dioxide (CO_2) in the environment is being absorbed at the ocean surface at a greater rate. This extra CO_2 results in increased hydrogen ions, which escalates ocean acidity.

Mitigation efforts to address global warming include the development and deployment of low carbon energy technologies, policies to reduce fossil fuel emissions, reforestation, forest preservation, as well as the development of potential climate engineering technologies. Societies and governments are also working to adapt to current and future global warming impacts, including improved coastline protection, better disaster management, and the development of more resistant crops.

- **Pollution:** It is the introduction of contaminants into the natural environment that causes adverse change. Pollution can take the form of chemical substances or energy, such as noise, heat, or light. Pollutants, the components of pollution, can be either foreign substances/energies or naturally occurring contaminants. Pollution is often classed as point source or nonpoint source pollution. In 2015, pollution killed 9 million people in the world.

Major forms of pollution include: Air pollution, light pollution, littering, noise pollution, plastic pollution, soil contamination, radioactive contamination, thermal pollution, visual pollution, water pollution.

- **Environmental Degradation:** It is the deterioration of the environment through depletion of resources such as air, water, and soil; the destruction of ecosystems; habitat destruction; the extinction of wildlife; and pollution. It is defined as any change or disturbance to the environment perceived to be deleterious or undesirable. As indicated by the I=PAT equation, environmental impact (I) or degradation is caused by the combination of an already very large and increasing human population (P), continually increasing economic growth or per capita affluence (A), and the application of resource-depletion and polluting technology (T).

Environmental degradation is one of the ten threats officially cautioned by the High-level Panel on Threats, Challenges, and Change of the United Nations. The United Nations International Strategy for Disaster Reduction defines environmental degradation as "the reduction of the capacity of the environment to meet social and ecological objectives, and needs." Environmental degradation comes in many types. When natural habitats are destroyed or natural resources are depleted, the environment is degraded. Efforts to counteract this problem include environmental protection and environmental resources management.

- **Resource Depletion:** It is the consumption of a resource faster than it can be replenished. Natural resources are commonly divided between renewable resources and non-renewable resources (see also mineral resource classification). Use of either of these forms of resources beyond their rate of replacement is considered to be resource depletion. The value of a resource is a direct result of its availability in nature and the cost of extracting the resource, the more a resource is depleted the more the value of the resource increases. There are several types of resource depletion, the most known being: Aquifer depletion, deforestation, mining for fossil fuels and minerals, pollution or contamination of resources, slash-and-burn agricultural practices, Soil erosion, and overconsumption, excessive or unnecessary use of resources. Resource depletion is most commonly used in reference to farming, fishing, mining, water usage, and consumption of fossil fuels. The depletion of wildlife populations is called defaunation.

4

Topic: Environmental Impact Assessment (EIA)

CONTENTS

4.1. INTRODUCTION

Environmental Impact Assessment (EIA) is a procedure involving evaluation of the potential environmental effects of a projected project or development, considering interrelated socio-economic, cultural, and health effects, both valuable and adverse. The UNEP describes EIA as an instrument used to detect the ecosystem, social, and economic effects of a project before decision-making (Figure 4.1).

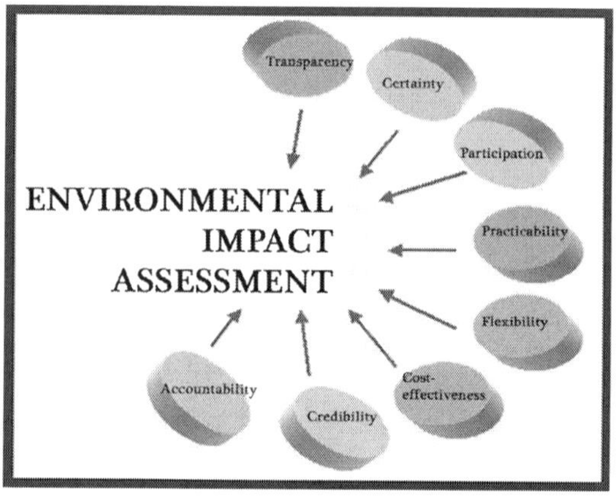

Figure 4.1: EIA is a multifaceted process that involves stakeholder participation, transparency, and flexibility among other aspects.

Source: https://www.pinterest.com/pin/758364024733124719/.

This process seeks to forecast environmental impacts from the earliest phase in a project's planning and design, discovering ways and techniques of reducing negative impacts, design projects to fit the local environment or present the estimates and possibilities to decision-makers (Wood, 2014).

Through applying EIA both environmental as well as economic advantages can be realized, like reduced rate and duration of project application and design, minimal treatment/clean-up charges, and impacts of legislation and regulations.

Even though laws and practices differ around the globe, the basic constituents of an EIA shall mostly involve the below stages:

- **Screening:** It's for determining which programs or developments need a complete or partial impact evaluation research.
- **Scoping:** This is meant to identify which possible impacts

are applicable to assess (reliant on legislative policies, global conventions, professional knowledge and public participation), to locate alternative solutions which avoid, alleviate or compensate negative impacts on biodiversity (such as the choice of not going forward with development, discovering substitute designs or sites that evade the impacts, integrating safeguards in the development of projects, or offering compensation for negative impacts), and lastly to originate reference terms for the assessment impact.

- Assessment and appraisal of impacts and establishment of alternatives, this is to predict and detect the likely ecological impacts of a suggested project or establishment, including the comprehensive elaboration of alternatives.
- Reporting of environmental impact statement (EIS) or report, plus an environmental management plan (EMP), and non-complicated summary for the overall audience.
- Appraisal of the EIS, based on reference terms (scoping) as well as public (covering authority) participation.
- Decision-making concerning whether to endorse the project or otherwise, and under the particular conditions; plus
- Monitoring, regulation, enforcement, and ecological auditing. Monitor if the projected impacts and planned mitigation measures happen as described in the EMP. Corroborate the acceptance of proponents through the EMP, for ensuring that unexpected impacts or botched mitigation processes are detected and handled in a timely manner.

4.2. STRATEGIC ENVIRONMENTAL APPRAISAL

Strategic environmental assessment (SEA) or is defined as a formalized, methodical, and comprehensive procedure of identifying and assessing the environmental effects of suggested policies, plans, and programs to guarantee that they're wholly included and properly addressed at the first possible phase of decision-making at par with financial and social concerns.

Because this early description of SEA has quickly developed and grown, plus the amount of SEA definitions has multiplied consequently. SEA, by its essence, now covers a broader variety of activities and a wider scope and usually across a longer time scale compared to the environmental

(EIA) impact assessment of programs. Furthermore, SEA may be employed to a whole sector (like a universal strategy on energy for instance) or to a terrestrial area (for instance, in the framework of a local development scheme) (Wood, 2014).

The SEA doesn't replace or limit the requirement for project-range EIA (albeit for some cases); however, it can assist with streamlining and focusing the integration of environmental concerns (such as biodiversity) right into the decision-making procedure, frequently making project-range EIA a more practical process. SEA is typically defined as being proactive and 'sustainability powered,' whereas EIA is normally defined as being mainly reactive.

4.3. PURPOSE OF ENVIRONMENTAL ASSESSMENT (EA) PROCESS

The goal of EIA is informing decision-makers plus the public concerning environmental impacts of applying a proposed project. Any EIA document by itself can be considered a technical tool which identifies, forecasts, and analyzes effects on the physical ecosystem, including social, cultural, as well as health impacts.

In case the EIA procedure is successful, it determines alternatives and moderation measures to limit the environmental effects of a proposed program. The EIA procedure also acts as a vital procedural system in the general decision-making procedure by promoting transparency as well as public involvement. It's essential to remember that the EIA procedure doesn't guarantee that any project shall be modified or prohibited if the process shows that there would be serious ecological impacts. As for some countries, the decision-maker might, in fact, choose the most environmentally-destructive alternative, so long as consequences have fully been revealed in the EIA. Meaning, the EIA procedure ensures that an informed decision is made, though not essentially an environmentally valuable decision (Wood, 2014).

4.4. PLANNING

Based on the EIA protocol, responsibility for generating an EIA shall be allocated to a particular party: (I) the state agency or ministry, as well as (II) the project proponent. In case the EIA laws allow, then either party might opt to hire some consultant to formulate the EIA or manage specific aspects of the EIA procedure, like public participation or practical studies.

Moreover, some EIA policies recognize the integral conflict of interest generated once a mining organization, or some other project stakeholder hires a consultant for purposes of preparing an EIA. Contracting a consultant has the risk that a document may be inclined in favor of ensuing with the project (Wathern, 2013).

In case a consultant has been hired by a mining company, it means conflicts might show if the consultant is confident they shall get future work by chance, the project is accepted. (For instance, consulting work for the port whereby ore shall be exported). A few laws need consultants to be listed with the state and/or officially accredited in EIA setup. For some cases, a consultant might be needed to file up a statement showcasing financial or some other concern in the aftermath of the project.

Preparing the final EIA: This stage delivers a final impact evaluation report which addresses the perspectives and viewpoints of the parties which obtained the draft EIA. The comments might encourage revisions or inclusions to the EIA draft. For some situations, the final EIA shall comprise an appendix which summarizes all comments obtained from the public including other interested bodies, and providing replies to those comments (Wathern, 2013).

Project Implementation: here, so long as all regulatory measures have been met plus permits are obtained, the project development will continue going by the project decision plus once prospects for administrative and/or legal review have been exhausted. Monitoring is also a crucial aspect of project implementation which serves three main purposes: (I) Ensuring that necessary mitigation measures have been implemented; (II) Determining whether mitigation processes are working efficiently; and (III) validating the precision of projections or models which were applied during the impact evaluation process.

4.5. RELEVANCE AND SIGNIFICANCE

Environmental assessment (EA) possibly weeds out environmentally-unfavorable projects; it further proposes modified designs for reducing environmental impacts not to mention identifying possible alternatives. In addition, it predicts significant negative impacts then attempts mitigation measures to lessen, offset, or abolish major impacts. Not to mention engaging and informing possibly affected communities and persons about the project. The assessment also impacts decision-making and creation of terms and conditions of a project.

4.6. EIA INPUTS TO THE PROJECT CYCLE AND THE OUTPUTS OF THE EIA PROCESS

4.6.1. Principles of Environmental Analysis

Scientists responsible for producing environmental data recognize the setbacks involved with obtaining meaningful data. Accurate data generation relies on planning, information collection, preservation, plus assessment of the sample. There's no particular aspect of this procedure that is far less significant than the next. For the data quality, it's generally as operational as the weakest linkage, the accurate analysis of ecological samples starts way before the ultimate analysis.

When more is learned concerning the toxicology of varying environmental analyses, consequently it becomes essential to monitor an ever-rising amount of ecological chemical species. Every analysis process needs its unique group of collection, analysis, and preservation methods which might or otherwise not be attuned with techniques applied for other comparable chemical species (Wathern, 2013).

Getting meaningful environmental data needs more than just a rapidly devised plan for collecting samples, filling a container, or dispatching samples to the lab; this needs a comprehensive knowledge of the system that's being sampled, as well as behavior of the object, and techniques accessible for analysis. The significance of a meaning and implication of the produced data can't be comprehended in data interpretation only, instead also covering the understanding of approaches applied to produce the data and having great impact on the ultimate result.

Environmental analysis mainly touches on ecological issues and progresses via the analytical methods that are presently used to acquire environmentally relevant information. The special relationship between different kinds of environmental samples, including the application of appropriate analytical methods necessary for obtaining useful data are also well captured (Wathern, 2013).

4.6.2. Environmental Categorization of Projects

As an aspect of the evaluation of environmental/social risks, including aspects of proposed investments, stakeholders apply the procedure of ecological and social categorization in reflecting the extent of risks and potential impacts. The ensuing category further specifies the stakeholders' institutional needs for disclosure based on their access to environmental data policy (Figure 4.2).

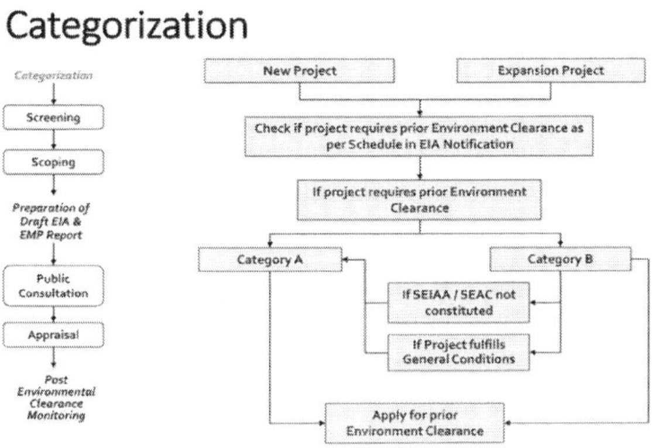

Figure 4.2: Project categorization can help with proper utilization of resources and better environmental management.

Source: https://slideplayer.com/slide/13637748/.

These classifications include:

- **Category A:** Industry activities with possible significant negative environmental or social hazards and/or effects that are varied, irrevocable, or unparalleled.
- **Category B:** Project activities with possible reduced adverse environmental/ social risks plus /or impacts which are less in number, normally site-specific, mainly reversible, and easily addressed using mitigation procedures.
- **Category C:** Project activities with reduced or no negative environmental or social hazards and/or effects.
- **Category FI:** Project activities involving cash investments in business institutions (FIs) or from delivery mechanisms that involve financial intermediation. The category is moreover subdivided into:

- **FI–1:** When an FI's present or projected portfolio comprises, or is estimated to have, considerable financial exposure to commercial activities with possible significant negative environmental or social effects that are varied, irreparable, or unprecedented.
- **FI–2:** When the FI's present or proposed portfolio consists of, or is projected to be consisting of, commercial activities which have possible limited negative natural or social risks which are lesser in number, generally site-specific, reversible, and easily addressed through mitigation processes; or covers a rather limited amount of business processes with possible significant negative environmental or social outcomes that are diverse, irrevocable, or unprecedented.
- **FI–3:** When an FI's present or proposed portfolio covers financial exposure to commercial activities that mainly have reduced or no negative environmental or social effects.

4.6.3. Sample Rapid Environmental Appraisal Checklist

The checklist is normally prepared for supporting the environmental sorting of a project. It's for attachment to the environmental classification form that's to be organized and submitted to a Chief Compliance (CCO) Officer of the Local and Sustainable Projects Department. The checklist must be completed with help from an Environment Expert in a Regional Sector.

The checklist emphasizes on environmental matters and concerns. In order to ensure social scopes are effectively factored, likewise, stakeholders must refer to ADB lists and handbooks concerning (a) involuntary resettlement, (b) local peoples planning, (c) poverty reduction, (d) participation, and (e) gender and development. These questions should be answered without ever assuming any "without mitigation" case. Generally, the goal is to detect potential impacts. Apply the "remarks" segment to discuss any projected mitigation procedures (Wathern, 2013).

Screening questions are also essential. It's important to check the potential ecological disturbances which may arise from establishing a plant, or facility complex within or around sensitive habitats. Some projects even cause eventual dilapidation of water bodies, because of discharge to wastes and other emissions from plant and facility complexes. Including serious pollution of soil and ground-water, worsening of solid waste cases in the region and public health risks associated with emission of wastes and low air quality.

An environmentally sound project should tick all of these boxes; otherwise, it would be a potential risk to the ecosystem. Projects are normally ordered during the initial screening of possible environmental impacts. Nevertheless, the categorization is subject to alteration as more detailed data becomes accessible and preparation ensues. Additionally, the environment category should be reconfirmed through the CCO before the original MRM process (Petts, 2009).

EIA is needed to examine the program's potential impacts, plus to suggest an environmentally sound program by comparing every likely alternative. Public consultation furthermore must be done at least twice throughout the EIA procedure, once during the initial phase of the EIA research and later following the preparation of a draft EIA report. Furthermore, the EIA must recommend appropriate mitigation procedures for limiting the adverse effects and determine environmental monitoring policies.

4.6.4. Environmental Assessment (EA)

It's the process of appraising and evaluating substantial short or long-term impacts of a project, based on the overall quality of its site's environment. EA also includes determining ways of minimizing, mitigating, and eliminating these effects or compensating for their impact. Also, EIA is normally prepared in the context of an EA, commonly known as environmental evaluation. EA is necessary for establishing the impacts, both negative and positive, concerning a particular project (Petts, 2009).

This will comprise technical evaluation, financial impact, and social outcomes that the project shall bring. In the U.S, it's the American Society of Testing and Materials that establishes the standard activities for environmental site valuations. The EA will need to do these particular steps: I) Identify potential environmental effects. II) Propose measurements to alleviate adverse effects. III) Predict if there would be substantial adverse environmental impacts, even following the mitigation implementation.

Additionally, environmental site evaluations aren't normally used to determine the nature and degree of contaminants found on a particular site. It's central to mention that environmental audits typically are applied to assess the environmental management as well as regulatory conformity of any given operation, but not via EAs (Petts, 2009).

- **Determining When it's Required:** EA is a multifaceted process that depends on various factors and variables in order to derive the most inclusive document. Typically, the design engineer shall determine whether you need it based on the magnitude and kind of work being planned. In other cases, the state may also need an environmental valuation to conform to both local and federal laws as well as regulations.

As for responsibility, it's the mandate of authorities to inform other agencies about their participation. They may also provide their expert evidence and data. Generally, the agency or body responsible for this study shall provide magnitude of the project; various factors that should be considered plus the anticipated timeframe for delivering the study are considered (ALEX, 2020).

Additionally, the environmental technician shall gather data; perform interviews and research concerning the impact zones. Thereafter, they would be able to detect the potential environmental impacts and measures for mitigating those effects.

4.6.5. Project Identification and Proposal

The role of project identification is developing a primary proposal for the highly appropriate mediations and sequences of action, within a particular time and budget structures, to address the particular development objective in a given region or setup. Furthermore, investment ideas may arise from various resources and contexts. They may come up from the location's sector policy, program, or strategy, as continuation of an existent program or from priorities defined in multi-stakeholder sectors or regional development dialogue.

Besides, identification includes:

- An evaluation of alternative techniques or options used for addressing a number of development issues and opportunities.
- The description of project goals and scope of task at the extent of detail appropriate for justifying obligation of resources for comprehensive formulation and corresponding preparatory research; plus.
- The detection of major issues which should be addressed and the individual questions to be answered before commencing the project, depending on the model should be applied.
- Adequate information on program options should be gathered to allow the authorities and financing bodies to choose a priority project, apart from reaching agreements among participants on

plans for work preparation, plus establishing steering committees and preparation teams for the project. Finally, results of the identification work must be summarized within a report, program brief or model document, the layout of which shall depend on the stakeholder's or financing body's requirements.

- **Why Identification Matters:** Proper identification is key to any project's success. If there's inadequate focus on projected outcomes or in case the perspective of the most practical concepts has been disregarded at the identification stage, there's little prospect which can be regained at a much later stage, when the concentration moves from assessment of options to explaining the particulars of a particular proposal. It may be expensive and hard to abort or fundamentally revise a model once preparation is ongoing (ALEX, 2020).

The burden to proceed quickly with project design, which may arise from both states and financing establishments, may lead to settling rapidly on a particular concept before adequate evidence has been provided to approve its validity. Various design-based problems experienced during application are the outcome of insufficient diagnosis of constraints, highly ambitious targets, time plans, and productivity forecasts, and inadequate examination of political buy-ins for the model and institutional volume matters.

Moreover, consultation, and participation of stakeholders during the identification phase is key to ensuring proper concept choice and increased potential for successful application and- eventually-sustainability. Model consideration must focus evidently on projected results and these must determine the proper interventions. Practically, concepts normally begin with ideas for mediations, and it's significant to rapidly direct the view on anticipated outcomes. Project ideas might arise from regional strategies or plans, emergent needs or outcomes from evidence-based inquiry, evaluation findings, as well as prospects for scaling up (Alex, 2020).

In many cases, professional bias is commonly seen in identification, since experts seem to focus more on solution paths associated with their specific area of concentration. Besides, the identification team must have the necessary expertise to comfortably identify and perform a primary screening of main alternatives. Additionally, close consultation with appropriate stakeholders in identification at both national and regional levels is vital for ensuring not just that the identification process draws on the top available data (evidence-based), but further that the growing concept will showcase stakeholder concerns as well as aspirations, plus that there's clear political and official buy-in to follow a project with an envisaged focus.

At the earliest stage, the formulation team must also launch contacts at the field stage with potential contributors and recipients of the projected investment, plus with the participants at community level. Allowing for active stakeholder engagement in the identification procedure will also inspire buy-ins, which improves prospects of application success and sustainability (Wathern, 2013).

The participatory workshop based on the results structure, using the log-frame planning technique or another similar technique is a useful technique for harnessing resources, including bringing together the perceptions of all stakeholders. The process can effectively outline stakeholders' needs, establish project goals and other associated project interventions, plus promotes comprehension of the allocation of implementation responsibilities. This may also be applied to detect knowledge gaps needing further inquiry and field survey, plus to evaluate likely resource accessibility and organizational capacity and community opinions.

Additionally, social analysis is highly useful when it feeds into an identification process, instead of being introduced at a later phase, since insights might have a decisive influence on priorities including the technique to be followed. For cases where thorough social analysis can't be done at identification due to time or even resource constraints, it means the team should draw upon the ideal and most appropriate existing analysis accessible, plus identify priority zones for more detailed inquiry during preparation (Wathern, 2013).

4.6.6. Project Appraisal

Appraising environmentally delicate projects must address various issues: including the political aspect of natural-resource programs, the capacity of environmental impacts, as well as the application of cost-benefit plus multi-criteria appraisal methods (Figure 4.3).

Process of Environmental Analysis

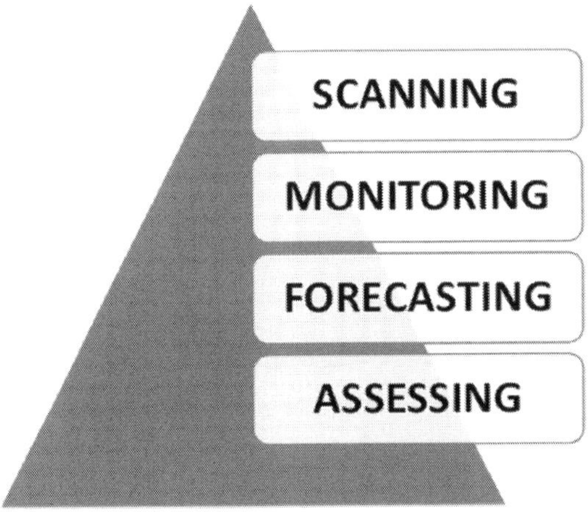

Figure 4.3: Appraisal looks at every angle needed to make the project viable.

Source: https://www.slideshare.net/zailunnito/environmental-appraisal-of-projects-79024560.

By definition, project appraisal refers to the procedure of assessing, structurally, the case for continuing with a program or proposal, and the project's viability. This normally involves comparing different options, through economic evaluation or some other verdict analysis method. The whole project must be objectively assessed for the same viability study must be done in its prime dimensions, technical, financial, social, and thus far to develop the project's justification. Furthermore, project appraisal involves judging if the program is profitable to clients or otherwise, including the detailed analysis of various aspects of a project before endorsement for application (Wathern, 2013).

Nowadays, drastic surges in human populations as well as per capita expenditure, climate change and overutilization of aquatic resources, among other factors have negatively impacted on biodiversity. These unwarranted situations are quite common and may have negative effects on the environment as well. Even though accurate valuations of threats have distinct importance for intervention planning, experts believe there's also the need to focus on successes. This reflection gives balance to negative situations and can swing attention towards productive, positive action.

Likewise, the nature of conservation efforts which the projects neighboring conservation community engages in can be perceived as an indicator of project appraisal success. To avoid possible termination, projects must address all possible environmental concerns before commencement. Stakeholders must discuss with the community all possible impacts of the project and allow them to give their input as well (Alex, 2020).

4.6.7. Project Implementation

Implementation basically means performing the activities defined in the work plan. It requires coordination of diverse activities, including the team's overseeing, management of budget, public communication and much more.

Regardless of what kind of project it is, there's a particular process that must be followed. Implementation provides the opportunity for watching the initial plans turn to reality. It allows end-users to gain access to improved outputs and living environments, not to mention success stories or experiences may further be shared with experts from other regions, therefore encouraging them to implement similar techniques, which consequently may enhance resources management within the surrounding region (Alex, 2020).

Project implementation or (or project execution), basically is the stage where visions or plans ultimately become realistic. It's the logical conclusion, following evaluation, determination, and visioning, planning, funds application and getting the financial assets needed for a project. Technical implementation covers areas like social development, environmental sustainability, health, and even capacity building. It's essential to take into consideration the uniqueness of the project's nature, some implementation processes take time, typically more than was originally planned, plus that various external constraints may appear, that ought to be regarded when initiating the application phase (meaning, seasonality in accessibility of community engagement (Wood, 2014).

- **Goals of the Implementation Stage:** The goal of implementation may be summarized as below. Setting into action established plans and operations. Before implementing any project, it's crucial that you develop a coherent and top-quality project proposal. Also, ensure that every possible role or responsibility has been well disseminated and understood. Both technical and non-technical work must be clearly described including the financial, technical, plus institutional contexts of the project, which must be prepared taking into account the local context.

Additionally, the working team must determine their potential strengths and weaknesses, prospects, and external threats. The strongpoints and opportunities are progressive forces that must be exploited to effectively execute a project. Weaknesses and threats, meanwhile, are hindrances which can prevent project application. Implementers must ensure that they develop a way of overcoming them. Yet another thing is that financial, physical, and human resources usually are readily accessible for usage (Wood, 2014).

Different other actions must also be undertaken before the task can commence, in order to execute a comprehensive action plan. Some of them include scheduling activities and detecting potential bottlenecks, networking with fellow team members and ensuring roles and responsibilities have well been disseminated and understood.

In addition, there must be realistic long-haul planning of finances, a process that's vital to the application of action plans. The communication strategy may be applied to raise awareness about the positive outcomes for the community, including mentioning that there are certain trade-offs, like the inclusion of pricing, which may not amuse everybody. It can help to further support local plan ownership and promote public participation for the execution of projects. By the end of the planning/implementation cycle, normally press releases are practical in highlighting successful stories and proclaim the final document's publication (Wood, 2014).

Likewise, expectations among stakeholders plus the general public normally are high after the participatory technique to the establishment of the preceding levels of the planning procedure. It's therefore essential that actions are noticeable and demonstrate concrete results early for building necessary confidence in the procedure. Based on the circumstance and complexity levels, a project may be applied through a formal production contract or even voluntary community technique.

Official written contracts are also needed where external contractors perform specialist development or installation effort. The work must be awarded via a competitive tender procedure to provide value for money. Other procurement strategies, furthermore, could be a designing and building scheme, or a build, own, operation, and transfer conception. Once official contracts are applied, there's a risk of omitting the social framework including the requirements of different users. Thus, it's essential to incorporate the community during this entire implementation stage, so as to encourage possession of the fresh infrastructure (Wood, 2014).

For community-managed projects, normally community members are active in the development and installation of fresh infrastructure through controlled labor agreements, personal contribution, work scheme foods, and individual-help programs where communities are offered training and resources for performing the work themselves. It's typically more affordable to utilize labor from around the community where the project is being implemented.

Nevertheless, there must be guarantee of quality plus a comprehension of the magnitude of the work. It's practical to have some written agreement between both primary and secondary shareholders (community and government) outlining roles and responsibilities, plus also approving the magnitude of the works. It will avoid unnecessary confusion or disagreements much later in the project's lifecycle.

It's crucial to remember that in both urban or semi-urban regions, many individuals already earn their living plus wouldn't be ready to contribute labor for a project scheme instead they would rather provide financial help. The situation must be evaluated at the earliest phases of a project. An effective approach for capacity development must be prepared to guarantee an effectual construction, operation, or maintenance work. Additionally, training activities shall target technicians, operators, masons, and different other service providers.

The activities seek to build the necessary capacities for the application and may be done through workshops, unique training programs, and "learning by doing" techniques, amongst other methods. No matter the technique that's chosen, the development and installation activities should be done under the regulation of professionals and project engineers (Wood, 2014).

The amount of supervision needed will normally rely on the context of the construction task. Nevertheless, if the design covers any engineering requirement, then qualified staff must be accessible on a full-time basis for overseeing the development of the works. Likewise, the overall quality of work may suffer in case supervision is insufficient, since corners may be expunged, substandard materials used, or safety compromised.

Similarly, the inclusion of the designers may further be required during the construction phase, especially to answer questions or make certain changes in the development whenever improvements or adaptations are needed. The practical application plan must be prepared with the implementation team for defining real-time roster for the delivery of outcomes (Alex, 2020).

4.6.8. Project Monitoring

Generally, a comprehensive EA program must be required of different projects, being part of their basic life cycle, plus the resultant data should be applied in environmentally responsible supervision and decision-making. This management may be aided through particular data on different environmental media, plus other natural and communal resource features. Present monitoring considerations are largely focused on the application of mitigation processes. Other valid reasons for environmental monitoring include, ensuring sustainability of the project for many years to come (Figure 4.4).

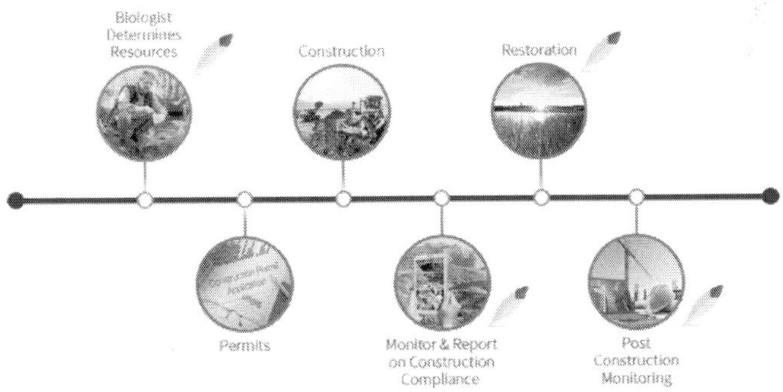

Figure 4.4: Monitoring ensures a project's conformity to set regulations and timely achievement of deliverables.

Source: https://wildnoteapp.com/environmental-compliance-monitoring-app-3/compliance-and-monitoring-environmental-project-timeline/.

An inclusive or directed environmental monitoring activity must apply extant monitoring information and coordinate with appropriate governmental monitoring structures. Additionally, program planning and application must delineate goals related to anticipated key impacts, choose pertinent indicators, decide sampling location plus frequency and analytical needs, predesign response strategies plus plan periodic reporting. Integration of environmental monitoring needs in the ecological impact assessment procedure would ideally need a mutual effort of stakeholders (Wathern, 2013).

Project monitoring has a significant role in project management and even the project manager's decision-making procedure. Nevertheless, it's a technique typically overlooked and only performed for the reason of accomplishing the needs of a project-management policy. However, when put into action, project monitoring may assist project managers including their teams to foresee possible risks and impediments that when left unaddressed, may derail the program.

It also clarifies the goals of the project, associate's activities to objectives, establish the target, reviews the project progress, and ensures the management remains aware of issues which develop during the application of the project. Apart from supporting and motivating the management to accomplish the program within budget, plus on time (Wathern, 2013).

- **Why Project Monitoring is Essential:** Whenever project managers make essential decisions but without confirmed data, it's like performing an activity in the dark. Decisions will be centered on minimal to no evidence at all, therefore the action might not be very effective and may just be nothing but waste of time or resources. Having a PMP certification shall help you to develop and become successful, especially as an object-oriented manager.

Therefore, it's significant to monitor projects carefully and apply the data gathered to develop intelligent decisions. Some questions that can be answered from project monitoring include; whether tasks are being performed as planned, the effect of these changes, or how your team is doing at any duration of time. Furthermore, automated equipment and technologies may simplify the monotonous procedure of project monitoring.

Many project managers have largely adopted project management implements to delegate errands and assess their projects. Nevertheless, project monitoring to a large extent is a complex procedure and there are just a few program management apps available that can promote the project manager's needs to have concentrated focus on unique tasks and team proficiency.

Project monitoring involves various steps such as staff meetings that can be done on a weekly, monthly, or yearly basis, statistics or advancement reports, and participatory evaluations by the stakeholders among other factors. For implementation, it's important to monitor the project planning parameter. This needs the evaluation of project parameters such as effort, schedule, costing, and timeline, among other parameters. It's the role of project managers to monitor these metrics while handling a project (Wathern, 2013).

4.6.9. Completion Report

A project completion report refers to a document that highlights the official closure of a project. Managers must prepare a project completion report to judge how it performed. It mentions what the program team delivered and evaluates budget and time performance. Since the role of a completion report is evaluating how the project performed, it's important to be candid and objective (Figure 4.5).

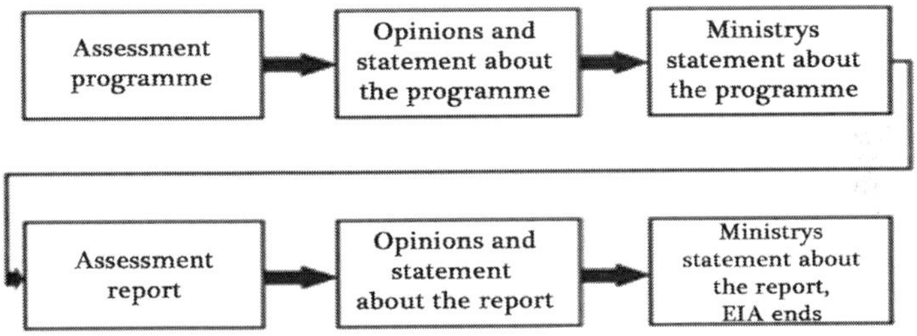

Figure 4.5: Completion report is among the final steps of a project that mentions its successes and possible failures.

Source: http://www.posiva.fi/en/final_disposal/nuclear_waste_management/permissions_and_procedures/environmental_impact_assessment_procedure.

- **Format:** Although the closure of a project report may take various formats, such as a formal exhibition, project management tool entry, or a document which may be passed about to different stakeholders, some of the below elements must be included within the project report:
- The title page with project title, including starting and end dates plus name of supportive and implementation agencies. This should be followed by a table of contents.
- Provide a summary of the project text, a brief statement that the assignment is complete should mark the starting of the overview. Furthermore, define your project at the problems background targeted by the project plus identify all goals and objectives that determine the project, including its intervention scope in the overview.
- Define the results and deliverables of the project. Similarly, add a section consisting of project highlights defining the most significant features of the project.

- Jot down the issues, possible challenges, and setbacks as a risk summary. Finally, mention the lessons learnt, activities that worked during the execution and what failed, including ways of enhancing the intervention. Likewise, you can provide some best practices at the appendix part of the project termination report.
- **Importance of End of Project Reports:** In case the project involves contribution from an external source, the project report might be needed by contract, particularly if the ratio or timing of the final payment would rely on the data in the report. Similarly, when the final report isn't particularly called for in the contract, both internal and external investors would most likely expect one, therefore it's ideal to deliver the report immediately after the project work is finished. The report may serve various purposes, such as:
- Notifying senior stakeholders, who might not have been fully engaged during any stage of the project, whereby the project is now complete.
- Notifying other departments or administrations that the work is done, plus that no extra resources shall be required for the project. It allows funding and workforce availability to be launched for use on different other projects.
- Recording any variations from the intended budget or schedule, alongside explanations on why the variations occurred. This may assist you to prepare a more precise project plan when managing similar tasks in the future, plus may also assist other project managers to organize their own projects.

Identifying the efforts of workers who contributed on the project, particularly those who contributed way more than what was required of them. This kind of formal recognition may really help towards building and preserving employee participation. There's absolutely no way of determining who may ultimately read your final project report, or whatever would be needed to reach it. It means your report must include sufficient data so that even somebody unaccustomed with the project may be able to comprehend the purpose, implementation, and the project's final result, without having to be overloaded with excess information (Wathern, 2013).

4.6.10. Evaluation

Evaluation is a key aspect of any project's cycle. It has two main dimensions which are: (a) self-assessment by the operations division responsible for organizing and executing projects, programs, and other operations; plus (b) independent assessment courtesy of local regulatory bodies. Self-evaluation involves the application of various instruments, such as (a) project/program performance accounts; and (b) review reports done during the period of project implementation, normally at midterm. Some of the essential tools for independent assessments of projects include the project performance assessment report, and program performance assessment report, respectively. The evaluations contribute to various other project studies as well.

To be beneficial, evaluations should be forward-looking (that is., determining issues, lessons, and endorsements that would be valuable for future operations) apart from assessing previous performance. The context for successful assessment work is done at the program, project, or TA design phase with the formulation of design and evaluation frameworks (Wathern, 2013).

Together with major performance assumptions, it shows a sequence of objectives, indicators of accomplishment with their targets, plus the techniques for evaluating the indicators. Some key features of design and monitoring structure constitute the framework of the program or project. Additionally, the performance management module, basically through the project performance reports, offers information required for self-evaluation and follow-up performance evaluation. Comprehending the political economy where the project is working is essential to evaluating the long-term sustainability, efficiency, and sustainability of the project.

Additionally, project ownership by stakeholders and beneficiaries is essential to guaranteeing project quality. Consequently, by way of participatory processes, the evaluation should also consider the perspectives of borrowers, implementing agencies, and recipients when evaluating the success or conversely of projects (Wathern, 2013).

It highlights the methods or significant issues to be assessed by the evaluation. this will normally include material about (a) how the borrower or beneficiaries will be covered; (b) the need for further research; (c) the subcategory to be applied in the ranking procedure; and (d) a re-statement, when necessary, a few of the impacts and consequence statements in the development and monitoring.

Finally, "impact" can be used to refer to a higher sequence, medium-range result where the outcome shall contribute, normally within 2 to 5 years of project termination. The output evaluation report must follow the structure mentioned in provided guidelines to guarantee consistency between assessments and the ease of tracing information in reports; nevertheless, minor deviations are possible to cover requirements (Wathern, 2013).

These guidelines are meant to help with analysis as well as report preparation. Moreover, they don't limit the obligation of evaluators to practice their best judgment, to prevent redundancies and replication, and to concentrate on significant matters. Besides, the report might quote openly from or offer cross-references for the appropriate project.

4.7. ENVIRONMENT IMPACT ASSESSMENT PRACTICES

4.7.1. Legal Framework for EIA

The main purpose of frameworks is to help project advocates, EIA experts, public members and lead sectors to comprehend and monitor the EIA process and for them to identify the phases at, plus the context where decisions on EIA implementations are made. It's hoped that this consequently will facilitate higher consideration and incorporation of environmental matters in development projects, plans, policies, and programs (Figure 4.6).

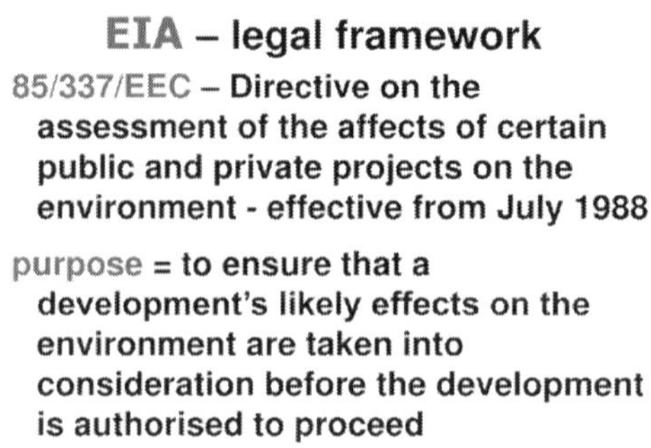

Figure 4.6: The EIA legal framework determines best practices for project implementation.

Source: https://www.slideserve.com/chars/environmental-impact-assessment-legal-framework.

Various sectors are needed to develop individual EIA procedures using the framework that national EIA Guidelines have developed. These frameworks are generally meant for members of the public, program proponents, learning centers, researchers, and policymakers. The procedures will be checked at various phases of the project sequence. Countries have since long pursued policies/strategies aimed at accomplishing reasonably great levels of development, for the rapidly growing population. Through time, natural resources have become quite stressed because of unsustainable resources utility, leading to shortages of essential environmental products in various parts of the world. It has made it essential to synchronize environmental laws for the reason of coordinating environmental administration (Whitelaw, 2012).

It involves the integration of environmental preservation in economic development for providing sustainable posterity development. This includes assimilation of ecological considerations in the development of planning at different levels; advancement of environmentally viable use of renewable/non-renewable resources for the procedure of national development; development of an institutional structure for coordination, monitoring, and enforcement of environmental protocols and standards; plus finally offering human and economic resources to support the environmental and development coordinating agency plus an EIA institution.

EIA Guidelines and Structural procedures are mainly developed in reaction to these policy frameworks and lawful provisions. Their key purpose is mainly to help in the assimilation of environmental matters to promote sustainable development. Furthermore, EIA identifies possible environmental effects of proposed developmental activities and policies, plans, and programs of the Authorities, including those done mutually with both bi-lateral/multilateral institutions. Additionally, EIA defines measures to alleviate the negative effects, while taking advantage of the positive policies. EIA is fundamentally a tool which enables knowledgeable decisions-making on viable development (Wood, 2014).

The government is ultimately accountable for issuing, varying, and possible cancellation of environmental impact appraisal licenses; they coordinate the entire EIA process. Furthermore, they are responsible for coordination of powers across various public/private sectors. Nevertheless, every sector plays some role in the application of EIA guidelines.

4.7.2. Project Approval Process

This involves decision-making at different levels plus the possible authorization shall be given when all EIA stipulations have been done and accepted by government and other relevant agencies. The EIA permit will be provided when the authorities are satisfied the EIA has satisfactorily been done for a project (Wood, 2014).

Additionally, authorities will offer a structure for conflict resolution in respect to: • Differences within and between State departments • Differences between State and Local Authorities • Disagreements involving public sector and private enterprise. Complaints concerning compliance with EIA stipulations which authorities may not resolve would be generally subject to review by an Environment Tribunal, having the provisions for bringing proceedings in the court of law wherever possible, for judicial review.

4.7.3. Legislative and Institutional Regimes for EIA

The purpose is to give an overview of various kinds of EIA protocols which are available. In order to determine the legislative, policy, and institutional provisions and directions that are available, it's recommended to familiarize with these institutional regimes before commencing with any project that has environmental impacts.

Relevance EIA occurs within the legislative and policy structures developed by individual countries plus global agencies. Besides, its practice may also be enhanced with a comprehensive understanding of the various arrangements which are provided for EIA provision as well as procedure, including how these may contribute to effective EIA (Wood, 2014).

Stakeholders developing or assessing EIA systems must be mostly aware of the positives and weaknesses of existing arrangements, plus the elements which can promote EIA as a device for achieving sustainable development. It must also be emphasized that EIA is more of an evolving procedure. During establishment and strengthening of EIA systems, normally there is an opportunity for taking advantage of the experiences of different stakeholders, plus moving towards both legal and policy systems that promote environmental sustainability.

EIA systems have increasingly become more widely based, covering an extensive variety of impacts, greater levels of decision-making plus new aspects of emphasis. Particularly, there are trends that are focused on: a) more systematic processes for EIA application, quality control, regulation, and enforcement; b) incorporated consideration of bio-physical, social, health, risk, and various other impacts.

Many lessons may be derived from this content by those that are accountable for introducing or adjusting EIA programs, or amending specific legal, policy or institutional structures to global standards. Not every aspect would be applicable or replicable in some developing nations, without additional EIA capacity development. Nevertheless, there's also a general trend for strengthening the fundamentals and key aspects of EIA frameworks in both established, and briefly determine the lawful and institutional features that are available in a particular country or locality and determine how they may have to be established or strengthened (Wood, 2014).

Each EIA system is unique to some extent, based on the political arrangement of a country. The EIA structure and modules from one nation (or global organization) might not be easily imported into another, even without considerable adaptation. The data obtained during the Analysis can help in detecting present and necessary activities in the strengthening of EIA systems.

There are certain key features that should be considered when examining EIA structures. It may be used to establish a profile of certain key provisions which apply, such as:

- designation of bodies responsible for managing the application of EIA process;
- the need for public participation, plus whether it's a compulsory or discretionary procedure; and
- procedural systems for EIA quality regulation, consisting of key levels of the EIA procedure (Wood, 2014).

Additionally, the matrix shall be most practical when applied to compare EIA structures of nations in the same area. Once complete, the table may be applied to detect directions where legal, policy, and organizational structures can further be strengthened. For some developing regions, for instance, the structures for public participation initiated by specific countries might vary considerably, showcasing various traditions and forms of governance. Likewise, some nations have set up different EIA authorities; for others the EIA procedure is overseen by the environmental department, or alternatively the planning counsel. There is no single EIA system that's appropriate for every country. Each country or region has its own unique system.

Ultimately, consideration may be offered to the extent where SEA or any near equivalent procedure is in place. Besides, an increasing ratio of developed nations and countries still in transition currently make official provision for environmental policies, plans, and programs. Lots of developing countries furthermore have planning systems, which include components of SEA. Additionally, the legal, policy, as well as institutional structures for SEA are widely varied, compared to those for EIA projects (Petts, 2009).

Also considered are universal developments which are relevant or appropriate to EIA's legal, policy, and institutional provisions for a country. Stakeholders identify features which may be applied in designing or establishing the EIA framework. The two main kinds of legal provisions which are developed for EIA include; (i) Overall environmental and resource management law, incorporating EIA requirements and protocols; and (ii) an EIA particular law, which may either be all-inclusive or assume the context of a framework or facilitating statute. Some examples of both national and global EIA systems; include the EIA modules and responsibilities which use 6 EIA Training Resource Guidebook, and Law, policy as well as institutional arrangements globally under some treaties or due to the lending needs of some key development banks.

Meanwhile, their geographical scale of application varies plus not all aspects would be applicable to certain countries. Some countrywide and international illustrations of EIA regulations. A good example of EIA legislation which establishes precedents or has been widely applied by other nations is the; U.S. National Environment Policy Act commonly known as NEPA. It's been dubbed the EIA Magna Carta. This protocol constitutes the foundational legislation of most environmental laws around the world, apart from being the pre-eminent declaration of the essence and reason behind EIA (Petts, 2009).

Additionally, the general language in purpose and declaration segments of NEPA matches with objectives and tenants of sustainability. NEPA was further used for the Rio Declaration for environment, particularly Section 102 which describes the procedural needs for EIS preparation. This has also been a subject of significant court re-interpretations.

Globally, the RMA is also used as a sustainability benchmark, and comes from a 4-year procedure of law and state reform. RMA is largely an omnibus law that was used to repeal or amend various statutes, protocols, and orders plus integrated their roles into a single legal regime using one purpose of 'promotion of sustainable managing of natural as well as physical resources.'

Section V of the Act outlines sustainable management amongst various other things like avoiding or remedying negative environmental impacts. Ultimately, this imposes some form of bio-physical test of sustenance on activities. RMA doesn't define the EIA process (rather this is described in a best practice guidelines provided by authorities) (Petts, 2009).

CEAA is similarly a model of an all-inclusive EIA specific law, which was passed in reaction to a number of legal challenges plus rulings on the earlier 1984 Guidelines Order. This legislation is of great interest globally since it entrenches the standard of public involvement, stipulates the roles of federal organs in regulations, plus prescribing the needs and procedure for applying various EIA levels. This Act applies just to projects; another separate SEA procedure applies to both policymaking and plans (Petts, 2009).

4.7.4. European Commission Directive on EIA

The EIA Directive generally is a charter law that's binding on affiliate states. It stipulates the basic principles and procedural frameworks for EIA in the EU, leaving things to the preference of member states on how these are converted into national legislation. Modern amendments to the 140 EIA Training Manual 2nd edition outlines the Directives, and have strengthened various key provisions, with particular emphasis on screening, reflection of alternatives, public discussion and decision-making. The proposed EC Directive for SEA on plans and programs are presently under negotiation as well, modeled under the EIA Directive.

The Directive requirements are furthermore captured in the EIA statute, policy, and institutional provisions of nations beyond the EU boundaries, particularly by applicant states of Central as well as Eastern Europe (CEE) that provide their individual EIA structures into line accordingly. Furthermore, the Directive may be anticipated to influence the EIA law-making process in other CEE states in transition plus might have a rather generalized impact as a reasonably standardized, widely accepted, minimum procedure for EIA.

4.7.5. Requirements for EIA

EIA is a crucial analysis that identifies both positive and negative impacts of a proposed developmental project, including how they affect people, individual property, and general environment. Besides, EIA also identifies certain measures to alleviate the negative effects, while maximizing on the positive ones. Basically, EIA is more of a preventive procedure. It aims to reduce adverse effects on the environment plus associated risks. In case a proper EIA has been performed, it means then that the environment's safety may be well managed at different levels of the project-planning, designing, construction, operation, assessment, and evaluation, including decommissioning (Petts, 2009).

The objective of EIA is ensuring that decisions on planned projects and activities become environmentally sustainable. It's performed so as to:

Determine impacts of the project on general environment b) Predict possible alterations on the environment due to the development c) Assess the effects of the different alternatives for the project d) Suggest mitigation measures for negative effects of the project towards the environment. e) Create baseline data for assessing and evaluating impacts, plus mitigation measures throughout the project lifecycle. f) Mention environment matters with the view to directing policymakers, stakeholders, planners, and state agencies to make ecologically and economically viable decisions.

- **Modules of EIA:** The main difference between Complete and Rapid EIA lies in the timeframe of the data provided. Rapid EIA is generally used for quicker appraisal processes. While both forms of EIA need inclusion/ coverage of various significant environmental effects and their mitigation, for the most part Rapid EIA realizes this through the assembling of 'one season' information just to minimize the time needed.

It is acceptable when it doesn't affect the decision-making quality. The assessment of Rapid EIA suggestions furthermore will reveal if a complete EIA is needed or not. It is therefore evident that the proposal of a professionally organized Comprehensive EIA for the first case would normally be the more effective approach. Based on the nature, locality, and scope of the program EIA report may comprise all or a number of these below components (Petts, 2009).

- **Air Environment:** This involves determination of the impact scope (via a screening model) plus developing an assessment network. It further covers monitoring of present status of ambient air-quality for an impacted area (7–10 kilometers from the margin) of the suggested project site.

It is also important to monitor the site-based meteorological data, wind speed or direction, moisture, ambient temperature and the environmental lapse rate. Approximation of ratios of air emissions comprising fugitive discharges from a proposed project must be considered. Also, identification, quantification, and assessment of other possible emissions (like vehicular traffic) around the impact area and approximation of cumulative of every possible emission/impact.

Stakeholders further need to predict possible variations in the ambient air-quality because of point, line, and zone source emissions, through proper air quality models. Assessment of the sufficiency of the projected pollution control instruments to meet gaseous discharge, and ambient air-quality standards are also necessary.

- **Noise Environment:** It involves monitoring the current status of noise amounts within the impact area and forecast of imminent noise levels coming from the suggested project and associated activities like increase in vehicular motion. Similarly, identification of effects due to some projected rise in noise amounts on the surrounding area. Finally, recommendations on possible mitigation measures to curb noise pollution.
- **Water Environment:** This is the study of present ground and surface aquatic resources regarding capacity and quality for the impact area of the suggested project. Likewise, prediction of effects on water resources because of the suggested water use/ pumping with regards to the project. Including quantification and characterization of wastewater covering toxic organics from the suggested activity.

The document is also focused on evaluation of suggested pollution prevention, plus wastewater treatment model and proposals on modification. If required-Forecast effects of effluent emissions on the context of the receiving water-body through appropriate arithmetical/simulation models—also, assessment of the viability of water recycling or reuse and delineation of a comprehensive plan in this context.

- **Biological Environment:** Survey of wildlife clearly defining season and duration. It covers assessment of wildlife present in the impact area of the project. As well as assessment of possible damage to land and aquatic wildlife because of ejection of effluents and gas discharges from the project. Furthermore, review of damage to both terrestrial flora/ fauna because of air pollution, land use, and possible landscape changes.

The biological angle further covers assessment of destruction to aquatic/ marine flora or fauna (covering commercial fishing) because of physical disturbances or alterations. Factoring also prediction of possible biological stresses around the project's impact zone, plus delineation of alleviation measures to avoid and/or lessen the damage.

- **Land Environment:** Here, the discussion is on studies of soil characteristics, existing land use/ topography, landscape plus drainage patterns around the impact zone. Including estimation of effects of project on general land use, topography, landscape, drainage, and hydrology. The identification of possible utility of treated waste in land application plus subsequent impacts. Also, estimation, and characterization of solid wastes plus delineation of management options to provide for minimization of waste as well as environmentally compatible disposal
- **Socioeconomic and Health Environment:** It involves gathering of demographic plus other related socio-economic information. The collection of epidemiological information, covering studies on noticeable endemic diseases (e.g., fluorosis, malnutrition, malaria, and fileria) plus morbidity ratios among the populace within the impact area.

It further covers projection of predicted changes in both socioeconomic and health because of the project, or related activities such as traffic congestion or delineation of actions to minimize negative impacts. Including evaluation of impact on considerable historical, cultural as well as archaeological sites/ places within the area, assessment of economic advantages coming from the project, and rehabilitation needs with particular prominence on scheduled zones, if any (Petts, 2009).

4.7.6. Environmental Risk Assessment

Environmental risk assessments are another crucial element of EIA involving hazard identification while taking recourse of hazard indices; catalog analysis and dam break possibility for water conservancy projects. Natural Hazard Probability is determined through a maximum credible accident (MCA) analysis for identifying possible hazardous situations. These include predicting the consequences of failures or accidents that result in fire, hazardous releases, explosion, and dam breaks among other factors (Petts, 2009).

Risk assessments also cover hazard and operability studies. An assessment of the risk-based on evaluations such as preparation of onsite/ offsite affected area, and disaster management plan or environment management plan. Also included is the limitation of mitigation measures, such as prevention and management for every environmental element and rehabilitation and resettling plan.

The procedure also covers the delineation of a monitoring scheme to provide compliance to enforcement conditions, and the delineation of the implementation plan covering scheduling and allocation of resources. Some common projects that are subject to EIA regulation include significant land-use changes, Urban Development and Transportation among others (Petts, 2009).

4.7.7. Institutional Aspects of Implementation of the Program

Successful project implementation ultimately relies on how every organization and stakeholder meets their responsibility in fulfilling the suggested aspects of the provided master plan. Nevertheless, when there's poor communication or co-ordination within or between organizations handling the project, this may prevent effective usage of available data and human or financial resources. Thus, master plans usually have a sixth module involving Institutional Plans for Implementation, this was suggested to advance coordination among different organizations and stakeholders, plus achieve incorporated management of projects, and the watershed provided in the primary plan.

The goal of institutional development may be summarized as; improving inter- and intra-organizational coordination, and clarifying environmental management roles to enhance the efficacy of environmental service delivery. Considering the scope of the government machinery, the development of yet another fresh body may not be necessary for public projects (ALEX, 2020).

Nevertheless, existing institutional arrangements aren't availing the amount of co-ordination or integration necessary for accurate environmental management and the subsequent watershed. For these circumstances, researchers propose the establishment of an organization called the "Conservancy," that is a recognized medium of stakeholders plus a body which can implement the overall resolutions of the forum. One of the key strategies for the Institutional Plan therefore is "establishing a Conservancy."

- **What's a Project Conservancy:** Experience over multiple years from different countries has inspired the creation of a 'conservancy,' being the institutional model which can be effective in management of multiple-use sites, especially those that have formerly experienced jurisdiction issues due to being within several administrative authorities. The classic 'conservancy' is managed by a board of stakeholder representatives (assisted, for instance, by provincial council, NGO representatives, municipal councils and the representatives from different conservation bodies.

Ideally, the model conservancy has a permanent staff with two key functions. First is providing the committee's secretariat, and next is implementation of committee executive. The conservancy therefore serves as both a representation body and managerial institution, with powers to make guidelines, collect fees and perform certain works (ALEX, 2020).

Additionally, the main benefit of stakeholders' committee is that it offers a common forum through which environmental issues, development proposals, and so on may be openly deliberated by the people's representatives who might have some interest in, or possibly influenced by, the project. Executive staff moreover is involved in various activities, such as, conservation/enforcement, environmental research/monitoring, and environmental protection/improvement just to mention a few (ALEX, 2020).

Basically, a conservancy is developed legally and can be funded through a number of: payments coming from regional authorities, grants given by central government bodies, and different research grants or international fund bodies. (A possible benefit is that when a legal institution has been developed, it may request for financial help from funding bodies, foundations, global donors, etc.) When established, the conservancy is identified as the approved guardian and 'voice' for the project, plus may be consulted therefore by developers, state agencies, and the public.

The inferences may be as follows: members agreeing with the formation of an effectual guidance and supervision body for the project. Besides, the proposed body must be supported by law and members believe this managerial body should have enough power for making decisions and executing them.

An independent body (quasi-governmental agency) must have an autonomous budget, so as to use every probable legal local and national fund. The new body should be able to use other financial resources in the law observation. For purposes of making and executing decisions, it's possible that the new managerial body must be consisting of representatives of different appropriate organizations and NGOs (ALEX, 2020).

Furthermore, it may also cover the use of management for the private sector. For the situation of urgent needs, the body may make use of various state and private organizations, this helps in making practical decisions. The management body shall comprise an independent formally determined system (the fresh body would need a committee plus an executive having a head and staff). Additionally, the execution of different affairs and projects tied to state and non-governmental organizations. All agreements will need to pass the essential legal steps. Following their acceptance, the Steering Committee teams shall be informed concerning the project conclusions.

Moreover, depending on the final result of further deliberations, the body may be a consultative group established at the provincial level, or even an independent body developed at national levels. In case the latter were chosen, the specifics of the fresh body may need to be considered, plus agreed upon during the period that the proposal is getting prepared for review by the Environment Council. Case in point, the particular roles of the conservancy will have to be deliberated. These may be restricted to the matters which all partners consent would ideally be supervised by the Conservancy. Any existing authorities wouldn't nevertheless lose their influence in the project (Alex, 2020).

4.8. FUTURE TRENDS IN EIA

In the coming years, the EIA Ordinance is expected to be reviewed to allow the public to contribute towards the EIA study scale, plus comment on possible EIA findings. Additionally, the law will formalize the consultation procedure with an autonomous Advisory Council for the Environment (ACE), which is selected by the State.

EPD is needed to give due concerning comments coming from the public, plus the ACE when possibly making decisions concerning EIA matters. Public participation means the public's involvement may influence decisions. Furthermore, with many years of positive EIA activities and experiences, the legitimate EIA structure has contributed towards environmental consciousness of various project proponents. Moreover, the extent may be reviewed in the EIA checkup, with an increasing cognizance of project proponent for adopting the suitable alternative from the environmental perspective, or adjusting the project for purposes of reducing environmental impact (Wood, 2014).

Case in point, EIAs for massive infrastructure projects, means the project proponents shall take into effect the existence of all significant environmental or ecological resources, plus choose alternative(s) for alignment in order to prevent and reduce effects to these zones. Through the documentation of deliberation of alternatives for the EIA report, both public stakeholders and other participants shall be able to appraise the selection procedure and provide their comments efficiently during public discussions.

Likewise, under the Environmental Impact Assessment Ordinance (EIAO) of Hong Kong, which was endorsed and promulgated for use worldwide. All proposed activities that are expected to have negative effects on the environment normally are considered as designated projects (DPs). EIA Ordinance furthermore offers the legal structure needed to apply the EIA protocol to DPs, plus make the necessary recommendations for the EIA legally implementable through the environmental permit (EP) (Wood, 2014).

Moreover, when a project proponent has already submitted the project profile, or EIA report, it means the EPD will assume the objective of supervising the EIA process, plus the general scoping and review. For the scoping level, EPD will furthermore decide the scale of the EIA document by providing the EIA Study Brief for the project advocate. Throughout the review stage, likewise, EPD will assess and make decisions concerning whether to endorse the EIA report which has been primed by the project protagonist.

5

Policies and Legal Aspect of Environmental Management

CONTENTS

5.1. INTRODUCTION

These refer to sets of laws that have been developed to protect the environment. The laws are commonly referred to as the laws that have been set and are required to be followed. They also refer to a set of procedures that are followed by the government or agencies that have the aim of protecting the environment. These laws/principles have been developed as a result of the degradation in the health and status of the environment. In the current day, air pollution, chemical spills, as well as harmful use of the land has become a common phenomenon. This has therefore caused a negative impact to the environment. These laws therefore seek to protect the scarce natural resources. For some systems of governments, they themselves have developed such laws by themselves while for other countries there are certain bodies that have been developed to make these laws and acts concerning environmental protection.

There are also some legislations that have been put in place that are also known as acts. Globally there are five effective parts of the environmental legislation. They include the clean air acts, the Montreal protocol, the reformation plan, the endangered species act, and the clean water act. For instance, the clean air act was developed as a result of most countries identifying that their air was having a bad smell and its condition was getting worse. This is attributed to the fact that most of the cities were said to be choking on smog. The clean air act therefore focuses on reducing reagents being released in the environment thereby reducing environmental pollution. It regulated the amount of carbon dioxide emissions. There is also the endangered species act which was first set out as a result of the animals such as the grizzly bear, red wolf, and the deer that have been endangered for a number of years. This act seeks out to ensure that such species of animals do not head to extinction (OECD, 2007).

Under this act, there is a classification that creates a distinction between threatened and endangered species. The endangered species are those that are at a risk of heading to extinction while the threatened ones are still present at a given number but are still hunted by poachers. These acts give a sort of protection to the two groups of animals. Some of these acts include the marine mammal protection that was signed in 1972 and its main role was to protect the marine ecosystem. The act is accompanied by a protocol that is followed. There is also the Montreal protocol act. This act is put in place to control the emission of CO_2. The rate of CO_2 to the atmosphere has reached high-levels such that one is able to get a sunburn when outside at a minimum of ten minutes (Figure 5.1).

Figure 5.1: The Montreal protocol act.

Source: https://www.airhawkac.com/montreal-protocol-and-the-phaseout-of-r22/.

This is due to the emissions of CFC molecules to the atmosphere that have contributed to the depletion of the ozone layer that has allowed harmful sun rays to penetrate and enter the earth. This act seeks to reduce the amount of such molecules being released to the atmosphere. There is also the clean water act that seeks to protect the waters more so the freshwater systems. It was noted that most of the rivers were turned into sewers. This is because of the heavy dumping that was done in rivers. Most of the rivers have been polluted by dumping plastics, wastes, and also oil spills in the rivers. This has also led to the killing and extinction of various life forms that have the freshwaters as their habitat. This act has undergone several amendments to control the release of toxic pollutants to the river (OECD, 2007).

The environmental policies vary depending on the problem that is at hand. For instance, countries facing much of water pollution implement more of the clean water act. Most of the environmental policies are inter-related to the energy policies. This is in that the energy production as well as the consumption will have its own effects on the environment. The energy production includes oil as well as natural gas. All the environmental policies seek to bring about a better management of the environment.

The environmental policies have changed according to the changes in the environment. If the problem worsens then the policies implemented are made to be stricter. Most of the environmental policies were put in place as early as the 1960s. For instance, in the 1970s the National Environmental Policy Act was established to monitor environmental quality. This body was later reformed to form the Environmental Protection Agency in 1971. This body made amendments to the clean air act as well as the clean water act. The body passed an endangered species act was developed in 1973. Such laws seek to monitor air and water. They also control pollutants such as industrial facilities and automobiles.

In the 1980s and 1990s, the act was allowed to manage and regulate contaminated wastes. They were also able to place some companies to be in charge of the contaminated sites. They manage the clean-up of such sites. It also led to the development of programs such as the superfund program that were established with the aim of cleaning the abandoned wastes sites. Most countries have a budget bill that allows the running of such programs. Before they enforce a bill, it is required that the bill is first submitted to the OMB before it is then affected. In the 2000s, it was proposed that a program be established to facilitate a market-based system allowing companies to purchase and sell pollution credits. With such regulations in place then it would have been a setback in the clean air act (OECD, 2007).

This program was not implemented as it did not progress beyond the congress. It aimed at reducing the greenhouse gas emissions. This was developed as a response to human-induced global warming. It was later that then the president allowed federal regulations that seek to reduce gas emissions from power plants and automobiles. It also included the signing of an act that allowed federal funding for the solar energy projects.

5.2. FEDERAL LEGISLATION

Here we get to further discuss the various acts. For instance, the clean air act was enforced by the congress in 1963 and was later amended in 1967 the latest amendment being in 1990. This act focused at minimizing the air pollutants as well as their sources of origin. The body that enforces this act is the EPA. It has laws that authorize the natural air quality standards. The clean water act was passed in 1972 by the congress. This act was an amendment from the federal water pollution act created in 1948. The body in charge of this act has the role of establishing the waste water standards. It also administers the national pollutant discharge elimination system. This system makes the discharge of a pollutant into waters without a federal permit illegal (Chaytor and Gray, 2013).

The endangered species act in 1973 was passed by the congress. This act gives a list of animals that are threatened and those that are endangered. It is composed of groups that enforce laws concerning the management of species under the conservation program. Under this act, there are outlines on actions to be taken on animals that are touched within the list provided (Figure 5.2).

Figure 5.2: The endangered species act in 1973.

Source: https://www.change.org/p/protect-the-endangered-species-act.

In 1980 there was the comprehensive environmental response, liability, and compensation act passed by the congress. It was established with the aim of minimizing the hazardous pollutants that are within abandoned or uncontrolled waste sites. Most of the waste sites were as a result of accidents, spills, or emergency release of some pollutants. This program is aimed at producing a list of these sites under a superfund program. The program is housed under the EPA. The EPA conducts and initiates cleanup projects in some of these sites. They also have other parties that pay up the costs for the cleaning projects.

The safe drinking act in 1974 was passed by the congress. It was later on amended in 1996. This act was made with the aim of maintaining the water quality. This is for both the underground and above groundwater sources. The body is in charge of setting the minimum standards for tap water. With the presence of such laws it is required that operators of public water systems meet the set standards. The amendment that was further made required that the EPA should conduct a cost-benefit analysis on drinking water standards. They are required to rely on the best scientific evidence available. The different state governments are given approval by the EPA to enforce federal drinking water rules. It is therefore the work of the federal governments to establish the minimum standards for drinking water (Chaytor and Gray, 2013).

In the United States, there is the department of the interior which is an executive department established in 1849. It was established to manage the natural resources within the state. It is in charge of the land agencies that give off the federal land policies. There is also a body in charge of the fish and wildlife service which seeks out to protect, conserve, and enhance wildlife, fish, and plants as well as their habitats. There are also programs that are in charge of the land management.

5.3. STOCKHOLM CONFERENCE

This conference was held by the united nation with the aim of placing more emphasis on the conservation of the environment. This conference was held in 1972. The united nation held the conference with the aim of providing practical means of encouraging the improvement of the human environment. It also gives off guidelines that the government and international organizations can take in a bid to improve the status of the environment. The conference was held in Sweden from June 5 to 16th. The conference was attended by delegates from 114 governments all having a common principle of leading their citizens in the enhancement and preservation of the environment. During the conference, there were some declarations or proclamations that were made (OECD, 2007).

Some of the proclamations were the protection and the improvement of the environment is an important issue that affects the good state of individuals as well as the economic development globally and is therefore the duty of all the governments. It was also noted that man has made use of their experiences as well as their discoveries. It was therefore declared that they use their creativity and skill in developing and enhancing their environment as it was noted that a variety of man's activities are leading to the destruction of the environment. This is attributed to the dangerous levels of pollutants in water as well as the disturbances in the ecological balance. It deals also with the disturbance of the ozone layer as a result of the release of toxic CFCs in the atmosphere. This has led to the depletion of some of the scarce resources that were intended.

It was also noted that in the developing countries, the challenges encountered in the improvement of the environment were as a result of lack of development in such countries. The resources in such countries though limited are not well utilized. In such countries, the level of human living decency is at a minimum. This is because most of the citizens are not able to access basic needs such as food, shelter, or clothing. It was declared that such nations were to make efforts in development that will be sensitive to the environment by ensuring that the environment is not affected. This is because in most of the developed countries the challenges that are encountered in terms of the environment are as a result of industrialization and the technological advancements (Welford, 2014).

It was also noted that with the continuous natural increase in the population, it will cause a burden to the environment. This is because most of the resources are limited and the increase in population increases the pressure of the environmental resources being able to meet the demand thus leading to the environmental exploitation. The preservation of the environment therefore becomes a challenge. As human beings are viewed to be more precious, it was declared that humans develop technologies as well as science that will be useful in improving the environment. It was noted that through ignorance and irresponsibility, massive destruction will be done on the environment. Most of the damages that will be inflicted will be irreversible.

It was declared that in order for the enhancement of the environment to work, knowledge, and wise actions will have to be employed. For the purpose of attaining freedom in the aspect of nature, people should use the knowledge they have to build and have collaboration with nature to enhance the environment. For this reason, goals had to be set. The goal was to preserve and improve the environment for the current and future generations. Therefore, fundamental goals of economic, social development and peace were laid out. In order to achieve this goal, it was declared that it is the responsibility of the citizens, communities, institutions, and enterprises to attain it. They all have to accept their responsibility. In every level, there should be equal distribution of roles so that each individual gets to contribute to the betterment of the environment.

The local and national governments are tasked with the role of improving the environment on a large scale basis by coming up with policies and actions that are to be enforced within the areas of their jurisdictions. In a bid to come up with the necessary resources to facilitate the conducting of the responsibilities, it was appointed that the international cooperation will take up the role of provision of such resources. The conference called out the people as well as the government to produce a general effort that will lead to the improvement and preservation of the environment as it will directly impact the people themselves facilitating their prosperity (Welford, 2014).

There were some principles that were also laid out during the conference. The first principle stated that man has a fundamental right to freedom, sufficient conditions to life and equity. All this is within an environment having quality allowing the dignity of life. Man is therefore tasked with the responsibility of ensuring the prosperity and improvement of the environment for the current and future generations. In line with this principle, oppressions and foreign domination are condemned and should be eradicated. The second principle states that all the resources of the earth should be safeguarded. The resources in this case include water, land, flora, air, and fauna of the natural ecosystems. They are to be conserved through proper management and proper planning.

The third principle states that the ability of the earth to give off new renewable resources should be maintained. If possible, this ability should be improved or restored. The fourth principle states that man is tasked with a special responsibility of managing and safeguarding the wildlife heritage as well as its habitat. It places more emphasis on nature conservation and the great need for planning for economic development. The fifth principle states that the non-renewable resources are to be utilized in a manner that will ensure they will not be exhausted. They should be present also for the future generation. They should also get the opportunity to make use of such resources (Welford, 2014).

The sixth principle outlines that the discharge of toxic substances and heat among others in the given concentrations should not exceed the environmental capacity so as to render them harmless. The discharge of such toxic substances was to be stopped in order to ensure that irreversible damages are not experienced in the ecosystems. This meant that countries struggling with high-levels of pollution should be assisted. The seventh principle focuses more on the water bodies whereby the various states should take all possible precautions to ensure the protection of the seas from substances that could in the future may end up being health hazards for human beings, damage to marine life and may cause disruption with the uses of the sea.

The following principle focuses on economic and social development. Emphasis is placed on the need for social and economic developments to facilitate favorable and good working environments for mankind. This will lead to the creation of conditions that will assist in the improvement of the quality of life. The ninth principle states that all the environmental deficiencies developed as a result of under-development or natural disasters produce problems that may be best solved through accelerated development. The accelerated development is achieved by a move on the substantial quantities of financial as well as technological assistance to supplement the domestic effort by the developing countries. For such to be conducted, timely assistance will be required (Welford, 2014).

Tenth principle states that for developing countries for environmental management to be possible it is required that there should be stability in the prices of commodities. Also, there should be adequate earnings for both primary commodities and raw materials. This is attributed to the fact that economic factors and economical processes are to be taken under consideration. The eleventh policy states that the environmental policies put in place by all the states should not affect the development of the potential of developing countries. It should instead try to enhance their development. They should champion for the attainment of good living conditions for every individual. The twelfth principle is more on the resources.

It states that resources should be at the disposal of individuals to allow the preservation and improvement of the environment. This involves taking into consideration that certain requirements in developing countries as well as the costs may result from the employment of environmental safeguards into the developing and planning. The thirteenth principle is that for a more rational management of resources to be attained the states should make use of integrated and coordinated approaches in their development planning. This is done to ensure that the development is aligned with the need to improve and protect the environment. This is done to benefit the environment (OECD, 2007).

The fourteenth principle deals with rational planning and it is composed of important tools used in reconciling any form of conflict between the need to improve and protect the environment and the need for development. The next principle states that planning should be employed in human settlements and urbanization in a bid to prevent harsh effects on the environment allowing maximum economic, social, and environmental benefits for all. This involves the abandonment of projects that facilitated colonial and racist domination. The sixteenth principle involves demographic policies that have no prejudice on human rights. They are deemed suitable by the government and must be applied in regions that have high rates of population growth that may impose harsh effects on the environment.

The seventeenth principle states that the most suitable national institutions should be given the task of planning, managing, or even controlling the environmental resources of states therefore enhancing the quality of the environment. The eighteenth principle makes use of science and technology where they are both considered to be part of the contributors to social and economic development. It is implemented in the identification, control, and avoidance of environmental risks. They also provide the required solutions to the present environmental problems and to also benefit mankind. The nineteenth principle deals more on environmental education. Environmental education is important as mankind will know of ways to protect and preserve the environment. This principle states that individuals both young and old should be given education on the management of the environment. It is required that mass media should be used to improve the environment by allowing communication on various matters pertaining to the protection of the environment. This is done by disseminating information on the basis of the educational nature of the environment and may involve the projection of things that need to be improved in managing the environment (OECD, 2007).

Principle 20 is no scientific development and research in the aspect of environmental management. These researches should be promoted more so in developing countries. This will facilitate the free flow of information considered to be up-to-date and hence the transfer of scientific information and information aimed at generating solutions to the current environmental problems. It also allows the transfer of environmental technologies to the developing countries and thus assisting them in improving their environmental conditions without necessarily creating a burden on their economic development. The twenty-first principle deals with the rights of the states. Here the states have to comply with the available charters of the United Nations.

The charters state that the states have the sovereign right to exploit the resources present in their own jurisdiction. This should be in line with the state's environmental policies. This ensures that there is control on the environment as well as the damages that would be imposed on the environment. The twenty-second principle requires the states to work together in a bid to come up with laws involving the liability and compensation for victims of pollution as well as environmental damages. The twenty-third principle states that a criterion should be agreed upon by the international community. This is also done alongside the standards that should be followed nationally. The laws are made with considerations to the system values of the given country. This is more so applicable for the advanced countries. If this is applied for the developing countries, then it would lead to the imposing of unwanted costs for such countries (OECD, 2007).

The twenty-fourth principles are on international matters involving the improvement and protection of the environment. It is required that such operations should be conducted using a cooperative spirit by all states. The cooperation may involve bilateral or multilateral agreements. It may also employ the use of suitable means to facilitate the prevention, control, reduction, and elimination of diverse environmental effects. Therefore, the sovereignty and interests of the involved states being taken to consideration. The twenty-fifth principle requires all the states to ensure that the international organizations to conduct a coordinated, efficient, and a dynamic role in the aim of improving and protecting the environment. The final principle states that humans as well as their environments should be spared from the effects of nuclear weapons as well as other means of mass destruction. Therefore, the involved states were required to make a prompt agreement on the elimination and total destruction of such kinds of weapons.

A conference is a form of a plenary meeting. The principles laid out are followed by the states that took part in the conference.

5.4. THE EARTH SUMMIT AGREEMENT

This summit took place in 1992 from 3rd to 14th June. It is also commonly referred to as the Rio summit. It was held in Rio de Janeiro, Brazil. The main purpose for this summit was to tackle issues concerning environmental protection as well as socio-economic development. This summit made an impact on the subsequent UN conferences. This summit research finding on the relationship between population, human rights, women, human settlements, and human rights. In the summit, there were a variety of conventions and treaties that were attained. They include the Agenda 21, Rio declaration on the environment and development, convention on biological diversity, framework convention on climate change and forests principles (Welford, 2014) (Figure 5.3).

Figure 5.3: Rio Summit, second earth summit (1992).

Source: https://sgerendask.com/en/summary-and-conclusions-of-the-rio-summit-second-earth-summit/.

During the summit, there were a variety of issues that were tackled. They include the issues pertaining to production of toxic components or poisonous wastes such as radioactive chemicals. The issues related to vehicle emissions, health problems as a result of air pollution and smog and the congestion in cities. They also discussed issues related to the use of fossil fuels that are in one way connected to the climatic change experienced globally. They also discussed issues related to the increased scarcity of water. The earth summit had relatively good attendance as 117 heads of states along with 178 representatives of the nations were in attendance. They all came together to rethink the advanced social equity, economic growth and ensure environmental protection globally. This was developed as a result of the Stockholm conference as it worked to improve some of the laws that were passed during the meeting.

The summit was created as a result of the response by member states to work together and work on the development issues after the cold war that caused massive destruction to the environment. After the war, issues such as sustainability were quite difficult for individual states to handle on their own. This led to the members of states resorting to work together in a bid to improve sustainability. This summit created a platform that allowed them to work together. As a result of the summit, it was noted that there was significant improvement in the sustainability of the nation's even to the non-governmental organizations. This led to other summits being held such as the case of the United Nations Conference on Sustainable Development being held in Rio in the year 2012. This summit was conducted from 13th to 22nd June and the issues addressed included the alternative sources of energy that would take the place of fossil fuels as they were noted to contribute to the global climatic change (Welford, 2014).

They also discussed issues involving the systematic patterns of production this more so laid focus on the production of toxic components and poisonous wastes. They discussed the increasing usage and limited water supply as well as the dependence on public transport systems as a solution to reducing the emission from vehicles. They also discussed health problems resulting from the air pollution and the congestions experienced in most of the cities. The major achievement realized was the summit reaching an agreement on the Climate Change Convention. This convention led to the Paris agreement and the Kyoto Protocol. They also reached an agreement that activities on the land belonging to indigenous individuals that would result in environmental degradation or one that would be considered to be culturally inappropriate should not be carried out.

There was also a convention on the biological diversity that was opened for signature by the summit. This led to the beginning of the redefinition of measures that would result in the reduction in the damaging of the natural ecoregions. This is because they will be required to not encourage activities that would bring about such kind of destruction which would result in uneconomic growth. It was later on noted that though the then president of the United States signed the convention on climate, it was quite difficult to negotiate its goals before the summit. This was one of the contributors that led to the failure in signing of the proposed convention on biological diversity.

The summit involved the awarding of nations that had innovative environmental programs. Twelve cities were given the award. Some of the cities include Canada and Sudbury that developed programs to facilitate the rehabilitation of the damages on the environment. Some of the damages were as a result of the mining industry. Some other programs include the green building strategy and the incorporation of an international education and training component to the municipal pollution control program. Among the documents that resulted from the earth summit is the Rio Declaration on environment and development. This document is composed of 27 principles that will guide the states in their future sustainable development. It is also composed of the precautionary principle which is principle 15 and the polluter pays principle which is principle 16 (Welford, 2014).

The polluter pays principle was enforced so that the party that led to the pollution will be responsible for paying for the damage done to the environment. This principle is regarded as a regional custom as it has received strong support by the European nations among other organizations. This principle has its own applications in the environmental laws by having different states adopting it. It also has international applications. The precautionary principle was a strategy developed and used in various approaches dealing with potential harm in areas where scientific knowledge cannot be applied. It places emphasis on caution, pausing, and review of innovations before leaping to them therefore preventing any potential disaster that would have been caused by the innovations. This principle faced criticism on the basis that it is unscientific and may be a limitation to progress. It is used by policymakers in events where there is a possibility of harm by making a certain decision.

There is also a document on the Framework convention on climate change. It was a treaty embraced on 9th may 1992. It was opened to signature in the earth summit. It was later enforced into law on 21st March 1994. This was a good number of countries that ratified it. This document was generated in an aim of stabilizing the greenhouse gas concentrations present in the atmosphere for it to attain levels that would avoid disastrous anthropogenic interference in the climate system. It was set out to create non-binding limits on the greenhouse gas emissions in individual countries. It did not have any enforcement mechanisms but it is composed of outlines on the specific international treaties that can be negotiated in an aim of meeting the organization's objectives (Welford, 2014).

The parties that met during this convention continued to meet annually since 1995. This was to allow them to make an analysis on the progress pertaining to climate change. It later on to the Kyoto protocol that was established and was a legal binding of the obligations for the developed countries. This required them to minimize the greenhouse gas within 2008 to 2012. It later led to the production of a document that sought to limit the future global warming below 2°C in relation to the industrial level. This agreement was later on amended but the amendment was not enforced contributing to the adoption of the Paris Agreement. This agreement focuses on governing the minimization of the emissions from 2020. This will be on the basis of the commitments of the countries in the Nationally Determined Countries. The agreement aimed at lowering the global temperature to 1.5°C.

There is also the United Nations Convention to Combat Desertification (UNCCD) document. This document was created for the countries that face serious levels of desertification and drought more so in Africa. It is a convention established with the aim of eradicating desertification and minimizing the effects of drought by making use of national action programs that use long-term strategies. The strategies are funded by international cooperation as well as partnership agreements. This convention was adopted in Paris on 17th June 1994. It was the only convention that sought out to address the issue of desertification. The convention was made on the basis of partnership, participation, and decentralization which form the backbone of good governance as well as sustainable development. The convention is made up of 197 parties meaning it tends towards universal reach. In a bid to make the convention to be known, its name was changed to the International Year of Deserts and Desertification (Kulkarni and Ramachandra, 2006).

This act is composed of national, regional, and sub-regional programs that have formed parts of key instruments used in the implementation of the convention. To strengthen it, it employs the use of action programs in both the regional and sub-regional levels. In order to involve the local communities in combating desertification, the national action programs framework was developed. This convention champions the use of science and technology being used to assist in combating desertification. It has allowed research to be conducted on the various ways this condition can be improved. There was also the Agenda 21 document which was an action plan by the US government in relation to sustainable development. This document is composed of actions that were to be carried out globally, nationally, and locally by the organizations in the UN. This document was later on implemented through the use of Rio+5 which was a special session by the UN general assembly.

To ascertain the compliance to the agreements made in Rio the delegates of the earth summit established the commission on sustainable developments which was later on replaced by the High-level Political Forum on Sustainable Development. This forum also meets annually and after four years as members of different meetings. It has been pointed out by critics that most of the agreements that were made in Rio have not been met more so on the policies that are affecting the fight against poverty as well as cleaning up of the environment (Kulkarni and Ramachandra, 2006).

5.5. ENVIRONMENTAL GOVERNANCE IN INDIA SINCE 1972

The environmental governance of India started after the then prime minister attended the UN Conference on human, environment, and development in 1972. It led to the formation of the National Environmental Planning and Coordination Committee. In 1972, there was also the setting up of the Central Pollution Control Board which was followed by the state boards. This later on led to the emergence of the department of environment. This saw the emergence of the environmental laws on water passed in 1974. The ones for air developed in 1981 and the forest conservation also in 1981. These laws were passed under the environment protection act in 1986. It also led to the formation of environment approval committees that were created for each sectoral assessment. India became a nation to follow the environmental governance system in 1996. It involved the further control notification of a coastal zone management, disposal of wastes and hill development.

However, things seem to shift to the reverse direction currently as there are no separate Union Ministers for the environment, climate change, and forest. It led to a 25% budget cut to facilitate the next step which involves MoEFCC. The government is ignoring responsibilities that will lead to the development as well as conservation of the environment. There were high expectations when bills such as the right to forest, right to food, right to information and right to 100 days of work were passed. The current impact felt is the regressive metamorphosis. The recent was as a result of the draft notification to amend the environmental impact assessment (EIA) that was to be done in 2014 and 2016. The amendment would lead to the increased power of the state governments. They will therefore be able to grant environmental clearance to development projects more so those concerning sectors such as river valley projects and mining (Chaytor and Gray, 2013).

In the 2016, amendment there was the formation of the environmental approval committees. By having the states being given the required power, it has led to changes in the areas of the river valley projects as large tracts of land have been covered. It was noted that the environment ministry had plans to move to the Supreme Court as they are against the National Green Tribunal having the real estate sector being exempt from environmental laws. There was later on a notification given that real estate projects up to 1, 50, and 00 square meters of built up areas make up the EIA. It was seen to be a poly that would allow the avoidance of the provision of environmental assessment (EA).

There was the generation of the green bench that stated the construction industry sector contributes to 22% of India's annual carbon dioxide emissions. This would affect the country's commitment to the Paris Protocol that aims at reducing the carbon dioxide emission that would cause a reduction in the global temperature. There were corporations that were given roles of forming the environmental committees that will be in charge of overseeing environmental aspects in line with real estate projects as well as construction buildings. Though these committees were made, critics argued that it would be questionable for these bodies to function without being influenced by the ruling political system.

It began by the de-notification of forests that are part of the declared protected area network. It was then followed by the country gradually leaning towards a single goal development at any cost. India is required to honor the 17 Sustainable development goals that were signed to the UN declaration. By doing so there would be breakthroughs experienced in environmental protection and improvement (Chaytor and Gray, 2013).

5.6. ENVIRONMENTAL PROTECTION AND FUNDAMENTAL RIGHTS

Environmental protection refers to a given proclamation of a human right to the environmental conditions to a given quality. The environmental protection and fundamental rights are interconnected in the rights cannot be enjoyed in the absence of a clean, healthy, and safe environment. Also, there cannot be sustainable environmental governance without the establishment and respect for human rights. Though there are such rights, it is noted that in a week three or more people may be killed while trying to protect the environment. Others end up being harassed, intimidated, or even criminalized. The fundamental rights have direct effects on the existence or enjoyment of the rights itself. They are composed of both civil and political rights, economic, and social rights, collective rights and cultural rights. Some of these rights include:

5.6.1. Right to a Wholesome Environment

It is also known as the right to a pollution-free environment. This right is present in article 48A of the Indian constitution. As humans have a very close interaction with the environment. They are therefore entitled to an environment that is healthy and by so doing will allow for sustainable development. Through the use of science and technology, there are certain methodologies generated to assist in maintaining the environment's health and condition. It is therefore the work of the government and the citizens as well to ensure that both the environment is pollution-free. With the increasing levels of pollution in terms of water and air, many individuals are being denied this right (Chaytor and Gray, 2013).

5.6.2. Right to Livelihood

This right is a principle that allows individuals to follow an honest occupation that will fully respect the natural world as well as other people. It would require individuals to take responsibility for the consequences of their actions as well as accessing a fair share of the resources available on earth. The right to livelihood allows individuals to uphold their principles and work in the face of diversity. It allows individuals to gain employment or work in a variety of areas in the aim of attaining financial stability. This will allow them to provide as well as sustain their environment. It allows individuals to live in peace, dignity, justice, and security. This right cuts across all walks of life as it is eligible for women, men, and children. This is under article 21 of India's constitution. This right is also eligible for foreigners that come into the country. This right may also house the right to freedom from sexual harassment while in the workplace. This right champions individuals being able to set up businesses by making use of the available resources on earth. However, the resources should be used properly to ensure that there are some of the resources left for the future generations.

5.6.3. Right to Equality

Equality means that every individual gets a fair of the resources. There should be no favoritism when allocating the resources. Both the rich and poor, young, and old should have an equal opportunity in accessing such resources. Not only does it apply for resources, it also applies in enjoyment of rights and fundamental freedoms. It is among the fundamental rights in India. It allows equity before law, discourages discrimination on various grounds such as race, religion, gender or caste. It houses equality in matters employment, titles, and abolition of untouchability. It therefore means that every individual will be treated equally. This is under article 14 of the Indian constitution and is used to ensure a guarantee that all individuals will be equally protected by the laws of the nation (Chaytor and Gray, 2013).

This law applies to all individuals that will be within the given jurisdiction and in this case, it is the Indian Territory. This means that also the foreigners and aliens in the nation are protected under this right. This law allows the poor to have equal chances of defending themselves when in front of a given court of law and thus allowing justice.

5.6.4. Freedom to Trade vis-à-vis Environment Protection

This has involved strategies such as free trade being used to foster environmental protection. This law was set up after it was discovered that trade has its own impact on environmental protection. It may either be positive or negative. Take the case of trade expansion leading to impact on the environment through the increased pollution and degradation in the natural resources. This act was developed under the American Free Trade Agreement. This agreement discusses various issues that deal with the relationship between trade and environment. This was as a result of the growth in trade that saw trade was rapidly growing to a point where it has reached to the international level. This is the matter that led to free trade. As trade involved the natural resources being used in conducting trade, it has led to exploitation of some resources. This led to emergence of trade groups that place some constraints on the use of the resources preventing them from being exploited.

5.7. A CASE STUDY OF INDIA

5.7.1. Ministry of Environment and Forest

This is an Indian government ministry. The headquarters are located in New Delhi. It plays the role of planning, coordinating, promoting, and overseeing the implementation of forestry and environmental programs. Some of the activities undertaken include conservation and survey of the fauna of India, flora of India, wilderness areas, and forests. They also work in ensuring the control and prevention of pollution, land degradation mitigation, and afforestation. They are also in charge of the administration of the national parks present in India. It is part of the All Indian Services. The organizations under this ministry include the India Forest Service that entails authorities, subordinate offices, centers of excellence, and autonomous institutions. So far, there have been a total of seven ministers who have headed this ministry. They have ensured that there has been a successful improvement of the environment.

5.7.2. Central Pollution Control Board

This board is under the ministry of environment, forest, and climate change. It was established in 1974 within the water act. It was given powers and functions under the air act. It plays the role of being a field formation. It also gives technical services to the ministry of environment and forests. It coordinates the activities of the state pollution boards by offering technical assistance as well as guidance. It also plays the role of resolving any conflict that may have arisen among them. This board is headed by a chairperson who is the career civil servant. The head offices are located in New Delhi. It is in charge of conducting research or EAs. It maintains the national standards in line with the environmental laws and local governments. It is required to conduct the monitoring of air and water quality. It also maintains the monitoring data.

This agency works alongside industries and governments in pollution prevention programs and energy conservation. It plays the role of giving the government advice concerning issues on controlling air and water pollution. They are also responsible for implementing legislation that is related to the prevention and control of environmental pollution. The board is composed of 500 full-time employees. The employees include scientists, engineers, and specialists in environmental protection. Some of the functions of this board in terms of air quality are to run some of the nationwide programs working to control air pollution. They are given frequent updates on the air quality updates weekly.

In the maintenance of water quality, where it is in charge of freshwater resources. They make use of the three-tier programs to check the quality of water through a network. They also have urban area networks that are used in urban areas. These programs are to ensure that the urban areas do not affect the environment. They also manage the municipal solid wastes. They are responsible for the collection, segregation, storage, transportation, processing, and disposal of the solid. The information concerning this is given to the authorities and used to provide technical assistance. They also come up with rules concerning noise pollution. They manage the various sources of noise such as industries, construction activities, industrial activities, loudspeakers, music systems, mechanical devices, and vehicular horns.

The board also manages all the information and data on the environmental statistics such as the water and air quality data. The air quality data may involve the levels of gases such as SO_2 and NO_2 among others. Water quality is in line with the quality levels of water in rivers and ponds. The board is composed of an organizational structure in charge of projects such as the pollution assessment, R, and D, laboratory management, training, development of standards as well as guidelines used in industry-specific emission and effluent standards. They also have an information database management and library; they make use of pollution control technology as well as pollution control enforcements. They conduct projects on mass awareness and publications with the aim of educating the public. They have projects on the hazard wastes management.

5.7.3. Environmental Protection Activities of the Department

Environmental protection refers to the protection of the environment. This is conducted by individuals, organizations, and governments all in the aim of conserving the natural resources as well as the existing natural environment. They also seek to make possible rehabilitation for areas that have been damaged in one way or the other. The damages are mostly as a result of the population growth and the overconsumption brought about by the ever-increasing population. This has been identified by the government and it has led to the government placing restraints on various activities that are considered to be contributors to environmental degradation. It has involved the creation of awareness on several environmental issues (Kulkarni and Ramachandra, 2006).

The environmental protection activities include having projects that will assist in the maintenance of the environment. They include tree planting activities, clean up in beaches and rivers, and also recycling of products such as plastics to avoid their congestion on the face of the planet. There are a variety of departments that head these activities and maybe developed according to the act in place. For instance, the act on water quality will require the team that is in charge of ensuring good water quality to come up with activities that will contribute to the cleaning of beaches. The departments get support from the government so that they can get funding to enforce the projects.

They may also get support from the international organizations who through the government may give funding for the projects. For most of these projects, they deal with the protection of the biosphere more so the ozone layer. It has been noted that in Africa the country embracing the biosphere protection is Tanzania. This country has enforced measures that have contributed to the good protection of their biosphere. Their natural environment is also well maintained (Kulkarni and Ramachandra, 2006).

5.7.4. The Statements of Objects and Reasons of Environmental Protection Act, 1986

This act was developed with the main aim of facilitating protection and improvement of the matters environment. It is said to be one of the most comprehensive legislations in relation to protection and enhancing of the environment. In the constitution of India, this act is under article 48A of the constitution which specifies that the state should strive to protect and improve the environment as well as to safeguard the forest and wildlife of the country. It is further noted in the article 51A that the citizens should also take part in the protection of the environment. It was noted that the environment was endangered as a result of man's economic activities. It makes use of a number of acts that will contribute to the protection and enhancement of the environment. The acts are designated according to the area of the environment that is affected.

The act was developed due to the visible signs of the decreased quality of the environment as a result of pollution in the environment. This act was signed in the Stockholm conference in 1972. The Indian government was among the governments that greatly voiced the environmental concerns. The laws on the environment currently focus more on the types of pollution or the certain types of hazardous chemicals that are present in the atmosphere. The act outlines the various measures that could be undertaken to ensure the protection of the environment (Kulkarni and Ramachandra, 2006).

5.7.5. Function of Government Agencies

The government agencies play a crucial role in the protection and improvement of the environment. The major government protection agency is the Environmental Protection Agency. Its major role is to administer the federal environment laws. It is also tasked with the role of converting the laws set by the congress into programs and regulations that can easily be understood by all who need to follow the outline laws. This agency distributes money to the states and certain organizations. This money is used as funding to run the environmental programs. It can also be used in the construction of sewage treatment plants. This will be a stepping stone in ensuring that the projects and programs will attain a level of success. This agency has employed over 17000 individuals including scientists, policy analysts, lawyers, economists, and engineers. Other countries have similar agenesis that work to improve the environment. To some extent, they are the ones to determine the projects to be undertaken. They also are in charge of forming committees.

5.7.6. The 42nd Amendment Act of 1976

This amendment was done in 1976 to the Indian constitution. This was done in an aim of creating a platform to allow the setting of protection policies. It was introduced to article 48A of the constitution. It read that the state shall strive to protect and improve the environment and to safeguard the wildlife and forests of the country. It also states that it is the responsibility of the citizens to contribute to the protection and improvement of the natural environment. The environmentalists greatly recognize the move by the government to improve the environment. The amendments were made together with article 21A which is the right to life.

5.7.7. The Indian Forest Act of 1927

The act was made on previous acts that were previously implemented while under the British. It was an act made in a bid to consolidate all the laws that are in relation to the forest. They also deal with the transit of forest produce as well as the duty that is placed on timber among other forest produce. The act contains the procedure that is to be followed when declaring an area as a reserved forest, village forest or a protected forest. It gives a clear definition on what a forest offence is as well as the acts that are prohibited while in the reserved forest. It outlines the penalties to be awarded to anyone who goes against the forest acts.

5.7.8. The Factories Act of 1948

This act was created to assist in the formulation of national policies with respect to the occupational safety and health in the factories in India. It was later on amended in 1987. The act therefore handles all issues dealing with the health, safety, efficiency, and wellbeing of individuals while in the workplace. The act is under the ministry of labor and employment. It is implemented in factories that make use of power and employs more than ten workers. There are various provisions that are within this act.

5.7.9. The Wildlife (Protection) Act of 1972

This is an act established by the Indian government to facilitate the protection of plants and animal species. It laid out the schedules of the protected plant and animal species meaning that the harvesting and hunting of these species is prohibited. It sought out to give protection to wild animals, plants, and birds. It also outlines the penalties that will be administered for those that breach the act.

5.7.10. The Water (Prevention and Control of Pollution) Act 1974

The act's major function was to ensure the maintenance and restoration of the wholesomeness of water by giving off a set of management guidelines and restrictions. It was added to the constitution of India in 1974. Some of its functions include the assessment of the water quality, pollution levels and also punish the polluters. The act also gives off guidelines in the aspect of disposal of wastes. It is composed of boards that will come up with rules pertaining to this sector.

5.7.11. The Water (Prevention and Control of Pollution) Act of 1977

This act sets out to provide the levy and collection of fees on recycled water that will be later on consumed by individuals carrying on given industries and local authorities. This is all with the view of augmenting the resources of the central board as well as the state boards in the prevention and control of pollution. This is all under the water prevention and control act of 1972.

5.7.12. The Forest (Conservation) Act of 1980

This act was established by the government of India to facilitate the conservation of forests and also manage all the issues pertaining to it. The act was later amended in 1988 and is applicable for the entire India. It was set to prevent further destruction of the forests of India. It is made up of five sections. The act was implemented on 25th October in 1980 and was later on amended in 1992.

5.7.13. The Air (Prevention and Control of Pollution) Act of 1982

It was set to facilitate the prevention, control, and reduction of air pollution by conducting for the aforesaid purposes. They are also given the mandate of allocating power to the various boards under it. It also lays out their functions. It was to bring about the control and prevention of air pollution in India. The act was later on amended in 1987.

5.7.14. The Environmental Protection Act of 1986

The act was set in the wake of the Bhopal tragedy. The act was signed under article 253 of the Indian constitution. Though it was passed in March of 1986, it was enforced on 19th November of the same year. It is composed of 4 chapters and 26 sections. Its main role was to implement decisions made by the UN concerning the human environment. Its role is to provide a framework for the central government coordination of activities established in the previous laws.

5.7.15. The Scope of the Act

The scope of the act is part of section two in the Act. It covers a variety of areas. They are composed of living organisms and inanimate elements of the environment. It covers both the natural and man-made environment. It thereof covers activities that may prove to be hazardous or may pollute the environment. It also covers the natural persons, unincorporated organizations and the legal entities. Under this, it discusses the rights and responsibilities that are in relation to the environment. It also outlines all the activities that can be conducted in relation to the environment. Finally, the scope also covers the performance of environmental tasks that have come about from international conventions. This is if the provisions of the international convention do not dispute it.

5.7.16. India's International Obligations

India is known for the large number of refugees it inhabits ever since they attained their independence. They have allowed their grounds to house refugees who have fled Pakistan. This has contributed to them encountering issues with their borders and refugees. This brought about the convention held in 1951 concerning the status of the refugees and later on led to the establishment of laws concerning the refugees. The large numbers of refugees would mean that the resources available in India may not be sufficient for all the population. India therefore had an international obligation of providing resources to all these individuals ensuring that they all have equal access to the natural resources. They also make use of resources obtained from other countries. They obtain all the resources from the UN. The international obligation also involves allowing foreigners to access their resources. This is under the guidance of the environmental laws laid down in the constitution. This is all in the aim of ensuring that the natural resources are not depleted. The resources should be maintained so that the current and future generations are able to make use of these resources. Even with the rampant industrialization in India by foreigners, they should make sure that their water pollution is well managed so that it does not damage the environment (Chaytor and Gray, 2013).

5.7.17. Public Interest Litigation

This interest is under the judiciary system in India. The public litigation meant that individuals that are not directly involved in the case presented in court being able to bring the notice of the court to the public interest. The court is therefore required to entertain the application for the public interest litigation. This method is good as it ensures that justice can be offered to the poor and it can help improve the weaker sections of the society. There are various considerations under this law. For instance, it was noted that the actions put in place for achieving the goals laid out in the Preamble to the constitution were limited. This meant that the benefits of the welfare measures stated in the legislation were not accessible to millions of the population. The attempts to create a bridge between the rich and poor did not produce the required results.

This also meant that the distribution of resources was not equal for all individuals. This goes against most of the acts laid out in the constitution more so those concerning the environment. The laws include the right to equality in terms of sharing of resources. The public interest legislation was laid out to improve this situation. The debate on this legislation was generated when the judiciary was giving its judgment on a case concerning sewage workers. It was realized that in as much as the judgment was to uphold rights such as the right to life, equality, and liberty the rights were not upheld for the poor as they were illiterate, ignorant, and were handicapped to poverty. This is because the Supreme Court did not entertain most of the cases that were related to the poor. This legislation championed for the rights of the voiceless to be heard before the judiciary system (Chaytor and Gray, 2013).

The legislation involves a procedure or an outline that could be followed in the case where the involved individuals feel that they were not granted the justice they wanted. There is a hierarchy in the judicial system that is followed in this kind of case scenario.

5.7.18. Public Participation

This gives an outline of how the public can work alongside the government in ensuring that they protect and improve the environment. The environmental management acts involve the public being involved in the various activities that will contribute to the improvement and protection of the environment. The boards involved in environmental management such as the EIA have the responsibility of giving the public knowledge on environmental management. They inform them concerning the various projects that have been proposed to improve the environment. They also allow the public to give feedback concerning the proposed projects. They also allow the public to give suggestions on how to improve the environment (Kulkarni and Ramachandra, 2006). The public is involved in all decision making activities that play an important role in the improvement of the environment. The public is involved in the stages of planning. The involvement of the public in activities involving the environment helps in assisting in increasing the public sensitivity and instilling personal responsibility to the public. It is one of the ways of giving environmental education to the public. This is done through conducting debates on the environment which by far has proven to foster active participation of the public in matters of the environment.

Public participation is aimed at minimizing the dependence of the government to make laws concerning the environment. Public participation has contributed to individuals coming up with groups or organizations that have assisted in improving the environment through activities such as tree planting and beach cleaning among others. Public participation is also among the acts under the environmental laws (Kulkarni and Ramachandra, 2006).

6

Environmental Management Systems and the ISO 14000

CONTENTS

6.1. INTRODUCTION

Environmental management is an organization of programs in a planned, documented, systematic way to reduce the negative impact on the environment caused not only by human beings but by animals or nature itself. Essentially a way a company organizes activities and sets strategies to coordinate the interaction, and more importantly, the impact of human society (or any negative impact thereof) on the environment. An organization develops Environmental management policies to strike a balance between meeting their needs and protecting the environment and Resources.

A Brief History of human beings managing their environment can take us back thousands of years to when Romans built aqueducts for water supply and medieval villages that learned to Channel their sewage away from town centers to prevent fouling of other resources. The evolution of environmental management systems (EMS) was mainly propelled by three ideas and developments

- The symbiotic relationship that we have with our environment;
- Improved health and science; and
- The foul nest syndrome.

The most considerable development in environmental management occurred within the 20th century. The early 19000 saw the first Air Pollution Studies and the first water quality test being carried out. Chlorination of water and other environmental laws were also carried out within this time period. When looking at water, for example, in 1900 800 deaths per year from water-borne diseases could be expected from a city hosting 1 million people. The first concept of managing the environment started from big cities like these where the full effects of fouling the nest were felt. A city like New York that was vast even in the early 20th century exhibited the negative impacts we had on the environment around us. To better, understand this evolution of and the rationality behind environmental systems of management we may divide it by the mediums the environment operates in; air, water, and soil, along with what constitutes as a negative effect on the environment.

6.2. HISTORY OF ENVIRONMENTAL MANAGEMENT

6.2.1. History of Air Pollution Management

Air pollution management involves our concerns on dirt dust soot and other pollutants and chemicals that have negative effects. Examples that can be raised are London's great smog of 1952 that made over 100 thousand people sick and killed 4000 people. Smog has been an issue in other areas of the US like Donora, PA, and several places in California. The term smog itself was coined from a contraction of fog and smoke but we should know it is much more complicated than that with other pollutants like nitrogen oxides and sulfur oxides being involved. The production of automobiles caused much of the smog but it was not the only source of tetraethyl lead pollution. Lead-based paint lead mining and smelting affected many parts of the US and the world over (Emblemsvåg and Bras, 2012) (Figure 6.1).

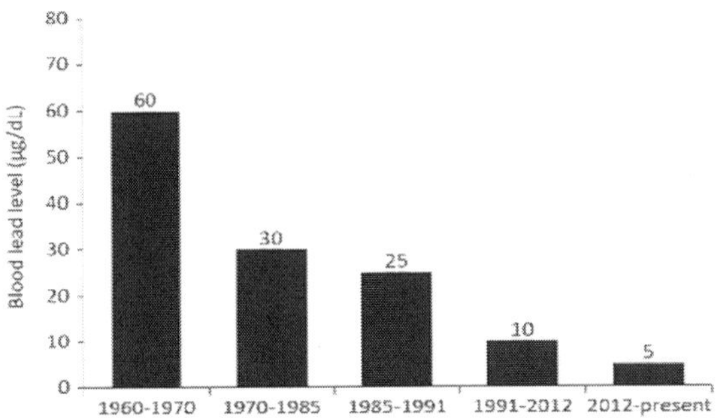

Figure 6.1: Lowering of BLLs by CDC over time.

Source: https://www.atsdr.cdc.gov/csem/csem.asp?csem=34&po=8.

The development of EMS that deals with air as a medium have been the most difficult to produce because of three main reasons. First is the difficulty measuring the form of Air Pollution as it is virtually ethereal in nature. Secondly is dealing with the medium itself. Is difficult to quantify and contain and a reasonable approach to gather as samples must be figured out.

The third reason is the type of pollution emissions in the air which come in three types' stationary mobile and Fugitive. Stationary sources of emission include things like smokestacks that are the easiest to measure. Mobile sources are constantly moving and include cars, trains, and such where sampling strategies are targeted along the highway, for example, to measure the general conditions. Fugitive emissions are the most difficult to track because they could come from virtually anywhere, for example, factories that have materials blown in the wind pollutants from construction sites of sewage treatment plants or lead paint flaking off the wall into the air (Emblemsvåg and Bras, 2012).

Mathematical models and predictions have been developed to measure air pollution as accurately as possible considering the aforementioned difficulties. Companies to be most concerned are those with high carbon emissions and methane like Saudi Aramco Chevron and Exxon Mobil.

6.2.2. History of Water Pollution Management

Between 1830 and 1850, 150,000 Americans died of cholera, a waterborne disease. The 11th president of the United States, James Knox Polk, died of the same disease. The beginning of the 20th century saw the first chlorination of drinking water and the main concern was how to control water pollution either by treating polluted water at plants that provide drinking water or controlling wastewater discharges. The main focus here was based on the dissolved oxygen in water and particulates, bacteria acidity, and nutrients that reside in the water. When dealing with wastewater when the sewage is dumped in a water source the naturally occurring microbes in the water feed on the sewage and oxygen is used. The oxygen is reintroduced back into the water in a process called re-aeration where the surface water absorbs oxygen from the adjacent overlying atmosphere. Many wastewater treatment plants still use this as a step in the water purification process and it is known as "natural purification." Where this naturally occurring process cannot happen, for example, when the rate of the degradation is higher than the rate of aeration, the water becomes toxic and putrid, nasty bacteria infest the water making its stink and fish die.

When the water has too many nutrients like phosphorus and nitrogen is eutrophicated. The excess nutrients in the water cause a large amount of algae to bloom, die, and decompose, swallowing all of the water's oxygen (Figure 6.2).

Figure 6.2: Toxic green algae in Copco Reservoir, Northern California (example of excessive eutrophication).

Source: https://www.nrdc.org/stories/water-pollution-everything-you-need-know.

Water pollution has many other angles besides just the ones mentioned above and over the last few decades, a better understanding has been developed outside of the conventional parameters. Priority pollutants, for example, are 126 pollutants that have high priority in the criteria of what determines clean water and was established in the clean water act of 1976. Mercury and acid rain are also a subset of water pollution concerns as Mercury is hard to eradicate and bioaccumulates in fish causing major health threats and does that consume and acid rain destroys whole ecological systems by dissolving essential minerals. The environment to management systems of waste management plants, water providers, geothermal plants, fisheries, and any other companies that relate with water as a resource are designed to combat the aforementioned water pollutants (Whitelaw, 2012).

6.2.3. History of Soil Pollution and Hazardous Materials

Hazardous waste is a relatively new phenomenon and it issues a new dilemma. Previously, it was automatically assumed that the best thing to do with hazardous material is to bury it deep underground and until very recently we did not understand that buried waste materials contaminate the soil and groundwater which provides almost 25% of our freshwater supply. The Love Canal incident, which initially started with the digging of canals on either side of the Niagara River that was abandoned and turned into a municipal industrial chemical dump site, was an unmitigated environmental disaster. The Hooker Chemical Company after many years of dumping corrosive material covered the whole canal with earth and sold it to the city. Torrential rain covered the area in 1978 and the love canal exploded and leaching began. Different pollutant compounds and suspected carcinogens percolated upward through the soil dumping contents into backyards of schools and hundreds of homes. Up until today, birth defects are recorded around the area with residents reporting a high-level of white blood cell counts which is a possible precursor to leukemia. Hazardous materials are very difficult to deal with in a soil simply because sometimes it is impossible to clean it up and the best measure is to make sure it does no further harm. Companies that must meet environmental management system policies on this front span from nuclear power plants all the way to hospitals that deal with bio-hazardous waste (Cassar, 2013). These environmental impacts listed above are some of the reasons why an environmental management system and other policies pertaining are very important as to reduce the negative environmental impact and also to improve the efficiency and operational effectiveness of a company or organization. A prescribed set of guidelines that can be constantly improved provides companies with actionable Solutions that they can depend on. The hooker chemical plant is an example of a company that did not have a well-established environmental management system to cater their waste products and by extent damaged their reputation and the environment of those around them (Figure 6.3).

Figure 6.3: Abandoned homes along the Love Canal, 1978.

Source: https://www.damninteresting.com/the-tragedy-of-the-love-canal/.

6.3. PRECURSORS TO THE ISO 14000

On a global scale, efforts have been combined over the last 40 years to meet and curb the effects that major companies and corporations have on the environment. The responsible care was started in Canada in 1985 as an initiative by the global chemical industry and runs in 67 countries covering 90% of the global chemical production. Companies that signed essentially pledged to occupational safety and health standards Environmental Protection and transparency with members of their publics. In 1992, the British Standards Institution developed and published the world's first environmental management system set of standards. This is key to mention because the same template was used to set up the ISO 14000 series in 1996 (Whitelaw, 2012).

6.4. THE EUROPEAN UNION'S ECO-MANAGEMENT AUDIT SCHEME (EMAS)

The British standards institution and their environmental management scheme was improved the year later in 1993 when the European Union prescribed tasks and responsibilities for member states in regards to their EMS. EMAS required that member states:

- Delegate their own independent competent bodies that are responsible for providing organizations with appropriate resources to meet environmental management standards.
- Appoint bodies that is capable of accrediting organizations when they meet environmental standards.
- Provide assistance to organizations and promote the eco-management and audit schemes (EMAS) to small organizations.
- Provide information about EMAS to the public and report back to the European Commission on progress in the implementation of standards.

Together with its predecessor the BS7750 of the British Standard Institution the EMAS integrated the provisions with the ISO 14001 standards. The latest EMAS update was passed in 2009 and post more stringent requirements for certification.

Before this, companies and organizations were not obligated but voluntarily created their own EMS. EMS were essentially the responsibility of a single office that monitored the regulatory compliance to minimize risk. This made it difficult to provide a benchmark that measured the environmental effects in which all these companies and organizations worked on, since every company had its own unique measurement standards (Whitelaw, 2012).

6.5. THE ISO

The International Organization for Standardization (ISO) is an independent, non-governmental organization created in 1947 in Geneva, Switzerland, and was intended to promote proprietorial standards, industrial standards, and commercial standards. It was initially set up in the 1920s as the International Federation of the National Standardizing Associations (ISA). As the world dove deeper into the Second World War it was suspended but afterwards was approached by the newly formed United Nations to form an acceptable global standards body (Whitelaw, 2012). It currently functions in 164 countries. Within it, over 20,000 standards have been set providing global standardization on everything from manufactured products and technology to food safety regulations, agriculture, and health care.

They have three types of members:

- **Participating Members:** Countries that have their own representatives standards body and they're the only members with voting rights within the ISO.
- **Correspondent Members:** Countries that do not have their own representative standards body but do not participate in the promulgation or proclamation of new standards.
- **Subscriber Members:** Small economies that pay no membership fees and are allowed to follow the development of new standards (Figure 6.4).

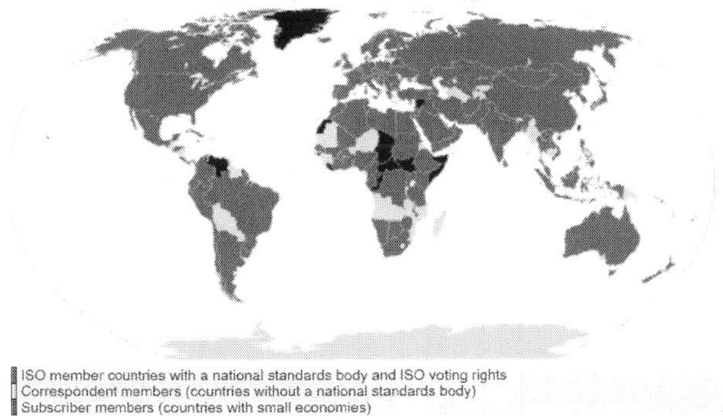

Figure 6.4: Membership of ISO.

Source: https://en.wikipedia.org/wiki/International_Organization_for_Standardization#Products_named_after_ISO.

The initial rationale on the creation of the standards was to develop international harmony especially with trade considering compliance to different environmental management standards across the world. ISO has continuously reflected the growing expectations that the public has on organizations to provide environmental credentials and to behave as the Agents of Change. Initially, the main focus was on pollution control and managing the negative impacts that organizations have on the environment but it has really grown beyond that

6.6. THE ISO 14000

The ISO 14000 is a family of standards specifically for the design, implementation, and optimization of EMS for businesses and organizations. 14000 is the number designated to this set family of standards on environmental regulations. According to ISO 300000, organizations across 171 nations are signatories of over 11000 rules. The first publication 14001 was published in late 1996 and provided detailed guidance on many if not all EMS related topics. Important to know that as one set of standards is revised which within the issue happens every 5 years the previous set of standards is withdrawn. This is because the framework set by the ISO has a continuous improvement approach to ensure that they continuously meet market requirements with a methodology known as the PDCA cycle (Morris, 2004).

6.7. THE PLAN-DO-CHECK-ACT CYCLE (PDCA)

This is a four-step model used for project planning that is meant to be repeated again and again for continuous improvement. The procedure is listed as follows

The first step is to plan where one recognizes an opportunity and organize a plan for change. Planning assists you with identifying all possible elements of the existing operation and therefore by extension future operations and how they may interact with the environment. Planning means considering processes resources responsibilities and coordination (Morris, 2004).

The second step is to carry out a small-scale study to test the change. In this step implementation of some of the processes, begin where the organization has already identified resources required and which members of the organizations are responsible for certain operations. This stage requires a high-level of efficient Communications for an effective EMS. The ISO 14001 implementation stage calls for:

- Operations control;
- Emergency preparedness;
- Documentation;
- Management of all resources (human natural financial and otherwise);
- Motivation for plan of action.

A successfully implemented EMS requires a full commitment from all employees beginning from the highest level of management. Sufficient resources must be provided for implementation and the traditional management culture must become environmentally-oriented discarding old practices and beliefs.

The third step involves the review and checking of results. Performance is to be monitored periodically to measure success in accordance to the organization's environmental targets and objectives. The correction stage takes compliance in relation to the objectives that were initially set by the organization. Some of the questions that can be raised in the process of reviewing are:

- Suitability of environmental policies;
- Recommendation to consider;
- New rules and regulations;

- Stakeholder interests;
- Public relations awareness and pressures.

The fourth step is to take action based on what has been learnt in the previous three steps implementing success and counteracting failures. Review in this state is particularly important in order to make recommendations for future improvements needed in the system and it should ideally be carried out by those at top management levels because their best positioned to assess effectiveness (Morris, 2004) (Figure 6.5).

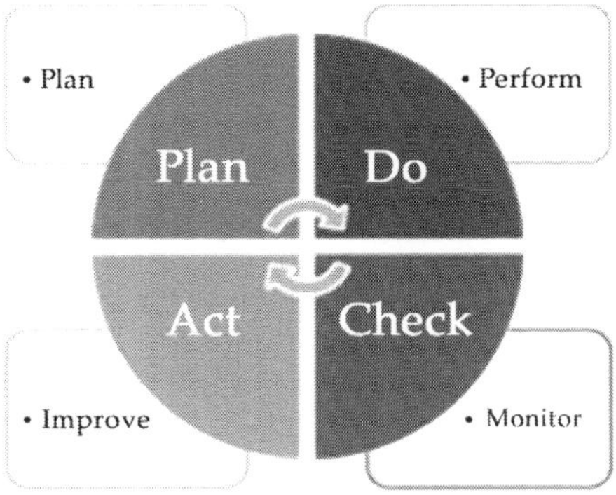

Figure 6.5: Plan-do-check-model.

Source: https://asq.org/quality-resources/pdca-cycle.

This cycle is used when:

- Beginning a new project;
- Developing new designs for a product or service and the processes thereof;
- Developing and defining a repetitive work process;
- Planning for the collection of data and analysis;
- Working on continuous improvement models;
- Implementing any change.

Two more important basic concepts/principles used by the ISO is the development of a good environmental policy and the continuous improvement process. An organization must be able to formulate and define an environmental policy. Here we may see clearly defined objectives and targets with principles along environmental sustainability. This policy should be clearly communicated both internally and externally and should describe an organization's goals and intent in regards to their environmental approach. The environmental policies are usually formulated by top management and it should be:

- Realistic;
- Well documented and available two employees and external parties;
- Clear on commitments to improving the existing EMS;
- Conforming to laws, rules, and regulations;
- Provide the framework to meet objectives.

The ISO further encourages an organization to ceaselessly make improvements on their environmental performance. The most obvious way to do this is by reducing the actual and possible negative environmental ramifications the organization may have. This encourages an organization to move away from basic operational environmental measures to thinking more critically and strategically on how to deal with environmental challenges. The six core elements, and this is in accordance to the ISO 14001 standards, are:

- Environmental policy;
- Planning;
- Implementation;
- Study and adjust;
- Management review;
- Continuous improvement.

The concept of continual improvement from an objective point of view is very attractive because it does not matter where an organization starts just as long as it starts secondly this also allows a company to build itself complementary to National regulations on environmental laws.

Because of the fact that the ISO 14000 series actually hold no environmental performance specifications that cooperation should adhere by there is a level of freedom that will also allow developing Nations to improve the environmental performance at an achievable rate while still being ISO 14000 certified (allowing them to compete with other nations and access to greater markets) (Whitelaw, 2012).

As much as it is a generic term referring to a whole set of standards it's cornerstone lies mainly in the specifications contained in ISO 14001 and ISO 14004 which will be explained in detail a little later but it is important to note there are two ways a company complies with them. This can be done either by self-proclamation by an internal audit or the EMS may be audited on a contractual basis by an outsider. To become officially certified and ISO 14000 however, a third party who is independent and nature must audit the company's environmental management system to evaluate its compliance with requirements.

6.8. BREAKDOWN OF THE ISO 14000

Probably the most important standard to highlight in ISO 14000 is ISO 14001 that details the specifications and guidelines of EMS to meet the requirements of a specific organization. Other key standards included in the ISO 14000 will also be discussed later.

6.9. ISO 14001 SPECIFICATIONS FOR ENVIRONMENTAL MANAGEMENT SYSTEM

The ISO 14001 is separated into 10 sections the first three pertaining to introductory information while the last seven list the essentials of developing a successfully functioning EMS. In total there are 16 standards six of which pertain to the organization itself and 10 directly to the products that they may produce again we must reiterate that it does not contain specific performance indicators but only establishes a framework for a company to build its own way of functioning.

Some sections in ISO 14001, like this section explaining compliance may appear to be deficient because it says very little about compliance itself, but it is not. The requirements say that an organization should have a policy that states in detail about their commitment being made to compliance and continual improvement. Essentially the organization must have its own procedure even to be registered to ISO 14001 standards. Once an organization has detailed its commitment through a policy, it must then create an environmental program to meet the targets and objectives of this policy and commitment. This environmental program has to be structured with a methodology at heart taking into consideration that the final objectives are not only the responsibility of the environmental department but goes all the way to top-level management. One requirement in this environmental program includes a periodic check on the overall compliance status which means that an audit must be conducted. Also described is a management review process as part of the compliance audit to perpetuate the concept of continuous improvement. Please is a good way for the organization to measure whether it needs to invest more resources or to make changes that would allow them to achieve compliance with the EMS (Whitelaw, 2012).

Because the ISO 14001 allows that voluntary standard to be sacked by organizations and provides the structure and tools to allow an organization to do this it means that organizations are now responsible and take ownership of their own environmental issues. This is done by an organization looking at its own systems and improving them to meet performance and Compliance goals rather than just focusing on improving performance continually

When success is defined by the organization itself top-level management is the best way to ensure continuous improvement and the ISO 14001 Fosters this with esprit de corps this helps further than the standing so that the organization itself through the sensitivity to the environment around them like never before (Whitelaw, 2012).

As discussed before the first three chapters of the ISO 14001 are introductory explaining the Scoop normative references and terms and definitions respectively, so the remaining seven sections are listed below:

- **Section 4 Context of the Organization:** In reality this section who's one of the most significant changes made in the ISO 14001 2015 standards. Context of an organization makes it necessary for organizations to consider the following:
- Direct environmental impacts of the organization.
- How the conditions of the environment impact the business, customers, and supply chain.
- External social-economic factors based on the organization's environmental performance.
- Internal capabilities of the organization to create change.

This is a lot to take into consideration and added significantly to the planning portion of the EMS to maintain ISO 14001 certification. Previously companies were only required to define the scope (activities, processes, etc.) in which they operate and build the EMS from there. The benefits of this enhanced approach are a better view of opportunities and risks to develop a better management strategy. This section is where auditors begin when assessing the EMS of an organization. Here the auditor will be looking at high-level issues as they are documented and how well commitments were made to comply with environmental regulations, reduce pollution, and continually improve.

Another change in this section includes the management's requirements to define the business environment pertaining to its internal and external environmental issues. These internal issues include things like air quality and waste disposal as well as the costs involved with external factors like legal requirements. From here, the organization is able to tweak its EMS to meet top management needs by undertaking environmental program activities like recycling and reduction of fossil fuel use (Whitelaw, 2012) (Figure 6.6).

	Organisational characteristic	Environmental condition
Internal Issues	Leadership support Resource availability Key skills Technologies, etc	On site: pollution, water quality, contaminated land, waste disposal
External Issues	Sector standards Corporate standards Legal requirements	Climate change Weather Floods Seismic events

Figure 6.6: Example of internal and external issues.

Source: https://isoconsultantkuwait.com/2019/06/22/iso-140012015-clause-4-context-of-the-organization/.

When thinking about the context of an organization there are many ways a company can have a positive, proactive impact on the environment around them depending on where they operate. Opportunities present themselves from improving the quality of the adjacent land or neighboring Waterways to reducing air emissions. A chemical plant or oil refinery may have many opportunities where in comparison to school may not.

Another key Focus in the IS0 14001 2015 edition is life cycle thinking, where consideration is taken into every step of a product or service from initial development to its end of life. At this point the organization can then consider creating its own compliance requirements, again voluntarily, assessing how they would impact the environment (Whitelaw, 2012).

This section has four sub-clauses:

- **Clause 4.1:** Understanding the organization and its context.
- **Clause 4.2:** Understanding the needs and expectations of interested parties.
- **Clause 4.3:** Determining the scope of the environmental management system.
- **Clause 4.4:** Environmental management system.

1. **Section 5 Leadership:** As already discussed above leadership is one of the top requirements of a successful EMS. Leaders are meant to define and communicate the environmental policy to their employees, assigning responsibilities and roles to each member to champion commitment to the EMS. In this section, guidance is given to an organization on how to demonstrate leadership commitment. They can be shown by:

- Paralleling the EMS with the organization's goals.
- Providing the necessary resources.
- Encouraging Active Performance to improve Environmental standards.
- Encouraging open dialogue with everyone involved in environmental matters and decisions.

Top management is responsible for creating a culture that encourages everyone towards achieving environmental targets by identifying opportunities to improve and participating actively in the EMS.

2. **Section 6 Planning:** Clause 6 of the ISO 14000 ensures that an organization is able to set the correct building blocks in place for the EMS to achieve the intended outcome. Here, top management must plan and assess where the risks and opportunities are and identify the environmental objectives improving on systems that were in place before. Risk and opportunity assessment are important parts of compliance and organizations need to think about not only the impact that they have on the environment but how the changing environment, especially resource scarcity and climate change will have on them.

In the ISO 14001 risks and opportunities is identified as potential adverse threats (risk) or potential beneficial effects (opportunities).

The sources of risks and opportunities are environmental aspects, compliance obligations, and issue of requirements from the contextual review.

- Risks turn into problems and we can reduce or even overall avoid problems by reducing consequences and probability, (which of the two parameters that define risk).
- Opportunities become benefits when the probability of the opportunities realized together with the value of the opportunity if it does become an eventual reality (the two parameters defining opportunities are value and probability).

A good example to offer is an organization working to improve energy efficiency measures which in the long run cut production costs (this is the value parameter) that creates the opportunity of the company being more competitive. Another example could be positive public relations surrounding the EMS (another value parameter) that would further the company's initiatives on gathering new customers (creating a new opportunity).

The clause has a further two sub-clauses listed below:

- Close 6.1 actions to address risks and opportunities. This section looks at the environmental aspects and Compliance obligations that the organization is to consider; and
- Close 6.2 environmental objectives and planning to achieve them. This clause concerned the planning actions then an organization to take to achieve the environmental objectives.

Essentially this section makes sure that an organization establishes implements and it continues to maintain (and improve) the processes needed to meet the requirements to address risk and opportunities and builds on the aforementioned understanding of the organization within the context its functions in (Morris, 2004). Some key factors to be stated here by the organization on integrating the business processes with environmental objectives are (Figure 6.7):

- What is actually going to be done?
- What resources are required?
- Who is responsible for meeting the objectives?
- When will these objectives be completed?
- How does the organization evaluate the results?

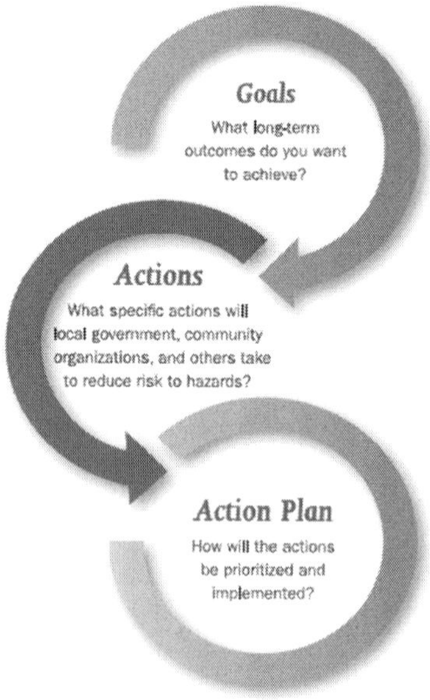

Figure 6.7: Framework to plan and EMS.

Source: https://isoconsultantkuwait.com/2019/06/22/iso-140012015-clause-6-planning/.

- **Section 7 Support:** This section is about the execution of plans that have been set by the organization to meet the EMS standards. Organizations need to determine which resources are required for the competence of the people working under the control of the organization in terms of their environmental performance and their awareness of the environmental policy. Because competence is a question here, there is also a criterion for EMS training where each person learns to play a role in environmental management. The company must ensure that training needs are identified and these planned needs are met. After this, a company is to verify the training has achieved its purpose whereby each individual is competent at applying this training and awareness to their particular job. Steps in developing an IMS training program are details as well.
- Step 1: assess training needs requirements
- Step 2: identify training objectives
- Step 3: choose the appropriate programs and methods

- Step 4: create a training plan
- Step 5: enforce the training program
- Step 6: build record that tracks training
- Step 7: evaluate the efficiency and effectiveness of the training
- Step 8: maintain and improve the training program as needed

The evaluation of training can be measured on several levels. Using employee interviews we can determine if the employees are content none of the training actually teachers the required skills and concepts. The IMS audit program can also assess whether the skills and concepts taught in training are used on the job and whether the application of these skills has a positive effect on the environment. Documentation also must have a way of showing that the employers understand the environmental policies in place (Morris, 2004).

- **Section 8 Operation:** The operation requirements are concerned with every aspect of environmental controls. The operation clause is divided into two sub-clauses:
- Clause 8.1 operation planning and control;
- Clause 8.2 emergency preparedness and response.

Clause 8.1 the design and reach of operational controls depends on the nature of the operations in the organizations and defined by the risks and opportunities presented with the aspects of compliance. Again, the organization decides the type of operation control methods to use. This can range from having processes that prevent error or ensure consistent results, using technology to prevent adverse results and having competent personnel to ensure the company meets CMS standards.

Many companies work by outsourcing products, services or processes where the ability has the organization to control please outside operation vary from direct control to limited or no influence. The ISO 14001 stator a good outsourced process is one that fulfills the below stated:

- It is within the scope of your organization EMS;
- It is important to the organization's functioning;
- It is needed it is needed to meet EMS standards;
- The organization and external provider have a relationship that is perceived by external parties as being carried out by the organization itself (Table 6.1).

Table 6.1: Example of Operation Control Measures

Processes	Significant Environmental Aspects (Considering a Life Cycle Perspectives)	Example of Controls (Considering a Life Cycle Perspectives)
Timber milling and sawing line • **Input**: raw timber and energy • **Output**: finished timber, cut to specification	• Timber sourcing • General energy use (petrol and electricity) • Temperature controlled storage (electricity) • Machine coolant use and disposal • Chemical use and disposal for wood treatment • Offcut and sawdust collection/ storage	• **Management:** Purchasing policy, custody chain certificates, procedures, and policies • **Operational:** Process flow charts, work instructions, signage, training, inspections • **Equipment:** Automation, designated storage areas, spill kits, metering, drain coding
Design office • **Input**: paper, inks, IT equipment, energy, customer specification • **Output**: design drawings and support material	• Procurement (Inks/paper/IT Equipment) • Energy use (Electricity) • Energy use (Gas-for space heating)	• **Management:** Purchasing policy, procedures, and policies • **Operational:** Process flow charts Training, IT procedures, draft reviews • **Equipment:** Default settings on printers, IT equipment, metering

Source: https://isoconsultantkuwait.com/2019/06/23/iso-140012015-clause-8-operation/.

Clause 8.2 is more concerned with emergency situations and how an organization is prepared to respond. Regardless of an organization's best efforts, Accidents and other emergency situations are still a possibility and effective planning reduces the negative effects of these emergencies by protecting employee's neighbors and interested parties. This emergency preparedness response should be reviewed regularly from continual improvement. The three components that they are so requires to be addressed are:

1. Establishing implementing and maintaining preparedness and response to potential emergencies this can be done by having a

plan that covers:

- The identification of critical assets;
- Describing the responsibilities to be taken up by the disaster team;
- Emergency services and enforcement services and how to contact them;
- Holding regular exercises.
- Creating a media Response Team responsible for contacting the media.

2. Continually review and revise to improve emergency response.
3. Periodically test the response actions were applicable.
4. **Section 9 Performance Evaluation:** This section is dedicated to measuring and evaluating your EMS to ensure it is effective and continually improving. It is important for an organization to know what to measure and the methods to analyze data and Report. Essentially an EMS without an evaluation process is like driving blind, you don't know where you are going. This clause is divided into three sub-clauses:

- Section 9.21 monitoring, measurement analysis and evaluation (which contains a very important section on evaluation of compliance);
- Section 9.2 internal audit; and
- Section 9.30 management.

The measurement section is again organization specific where it could include procedures like measuring the weight of waste sent to landfills, the amount of electricity or gas consumed per week or month or even measurements of noise pollution on site. These measurements help you monitor the environmental performance and analyze the root cause of your environmental problems.

Management theories have endorsed the use of the "vital few" concept when measuring and monitoring. The vital few consider a limited number of factors that would have a significant impact on outcome where we ask which operations and activities have the most substantial environmental impact, what are the characteristics of these activities and how can we measure them. When considering measurements systems, we can define two types of measurements (Figure 6.8):

- **Outcome Measurements:** Measurements of final results after process has been completed.
- **Process Measurements:** Measurements that happened while a process is ongoing.

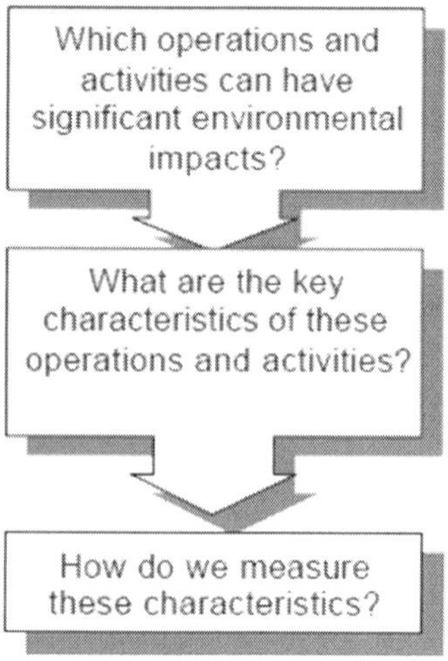

Figure 6.8: Vital few flow chart.

Source: https://isoconsultantkuwait.com/2019/06/23/iso-140012015-clause-9-performance-evaluation/.

These measurements track performance and based on the measurement criteria determines the overall success of the EMS program. Performance indicators give us qualitative data on our measurements. An example would be if a company measured a reduction of waste, it may only mean that the waste is being transferred through another medium and not actually reduced; so reduction should be measured from the source where the waste is generated for it to have any quantifiable meaning

The ISO 14001 states three types of environmental performance indicators (EPIs), namely:

- Management performance indicators (MPIs) that measure the effectiveness of the policy people activities and practices within the organization.
- Operational performance indicators (OPIs) that deals with input and output and the whole operational processes within the physical facilities.

- Environmental condition indicators (ECIs) that discuss the condition of the environment in relation to the organization and its impact thereof (Figure 6.9).

OPI	MPI	ECI
Raw material used per unit of product (Kg/unit)	Environmental costs or budget ($/yr)	Contaminant concentration in ambient air (μg/m^3)
Energy used annually per unit of product (MJ/1000 L product)	Percentage of environmental targets achieved (%)	Frequency of photochemical smog events (#/yr)
Energy conserved (MJ)	Number employees trained (% # trained/to be trained)	Contaminant concentration in ground- or surface water (mg/L)
Number of emergency events or unplanned shutdowns (#/yr)	Number of audit findings (#)	Change in groundwater level (m)
Average fuel consumption of vehicle fleet (L/100 Km)	Time spent to correct audit findings (person-hr)	Contaminant concentration in surface soil (mg/Kg)
Hazardous waste generated per unit of product (Kg/unit)	Time spent responding to environmental incidents (person-hr/yr)	Concentration of a contaminant in the tissue of a specific local specie (μg/Kg)
Emissions of specific pollutants to air (Ton CO_2/yr)	Number of complaints from public or employees (#/yr)	Population of an specific species within a defined area (#/m^2)
Wastewater discharged per unit of product (1000 L/unit)	Number of suppliers contacted about environ. mngment. (#/yr)	Fish deaths in a specific watercourse (#/yr)
Air emissions were exceeded (days/yr)	Management levels with specific environ responsabilities (#)	Employee blood lead levels (μg/100 mL)

Figure 6.9: Examples of performance indicators.

Source: https://isoconsultantkuwait.com/2019/06/23/iso-140012015-clause-9-performance-evaluation/.

Compliance obligations are defined by the ISO 14001 as "legal requirements that an organization has to comply with any other requirements that an organization has to or chooses to comply with." To put it to put it bluntly compliance obligations replace legal and or other requirements while never falling below them. Actions that address compliance obligations are fixed in the EMS and all other business processes and the organization are required to evaluate and document the level of compliance. The process of evaluating compliance must include frequency of compliance evaluation meaning that regular review and periodically checking how well you are performing is necessary and an organization is required to always know if they are reaching these levels. The elements of meeting these levels of compliance obligations are:

- Commitment to compliance;
- Determining compliance obligations;
- Converting compliance obligations into impact on the organization;
- Insuring measurements are taken;

- Self-assessing compliance;
- Internal audit and Management review of compliance.

Section 9.2 that discusses internal audit programs has always been an integral part of ISO 14001 to measure the effectiveness of an environmental management system. To ensure consistency objectives and scopes are to be defined for each audit in a thorough and specific fashion. Audits are defined by the ISO 14001 as "a systematic and documented verification process of objectively obtaining and evaluating evidence to determine whether an organization's environmental management system conforms to the system audit criteria environmental management set by the organization and for the communication of the results of this process to management" the auditor must know exactly what to measure and management then acts on feedback. A company should be serious on following through on the improvements and they will surely see the benefits of this (Whitelaw, 2012). An audit program that is effective should:

- Develop audit protocols;
- Establish proper audit frequency;
- Train auditors;
- Maintain audit records.

Audit methodology listed in the ISO 14001 or compliance auditing and Performance auditing. Compliance auditing is the traditional Style that is very procedural, taking measurements, updating legislative registry and generating proper records. If one meets the requirement of ISO 9001 such auditing methodology is very straightforward as it is very similar to quality assurance systems auditing and if your company is registered under ISO 9000 you can consider using your internal ISO 9000 auditors as your EMS auditors (Whitelaw, 2012).

Section 9.3 discusses management review where top management systems are held responsible for reviewing the EMS for its continuing improvement adequacy and effectiveness. This environmental management review is not detailed in terms of how to perform it and it should be tailor suited to the organization's culture and resources. Adequate communication channels for employee consultation must be maintained.

- **Section 10 Improvement:** This clause is related to organizations identifying opportunities for improvement of the environmental management system taking into account the continual improvement factors.

Some of the internal factors influencing improvement of the EMS include:

- And levels from top management on environmental policies;
- Level of involvement of employees;
- Level of training of the employees;
- Process design;
- Supplier management;
- Design of measurement;
- Information management.

The external factors influencing improvements of the EMS include:

- Competition pressure;
- Customer and social complaints;
- Market requirements;
- Financial support;
- Legal requirements;
- Requirements of certification bodies;
- Corporate requirements.

The first section of Class 10 improvement is pertaining to general information and the other to discuss nonconformity and corrective action (Clause 10.2) and continual improvement (Clause 10.3) no environmental management system is perfect and organizations are to establish and maintain procedures 4 non-conformities as the organization and the EMS will need to change as it grows. Major and minor non-conformities are dependent on the type of organization that is operating. A nuclear power plant, for example, has little room for non-conformities than car manufacturing plants.

Section 10.3 deals with continual improvement. It is important to note that this is a concept that the ISO 14001 stated as a long-term objective.

This marks the review of ISO 14001 in a nutshell. Other standards in the ISO 14000 family that are most notable include (Figure 6.10).

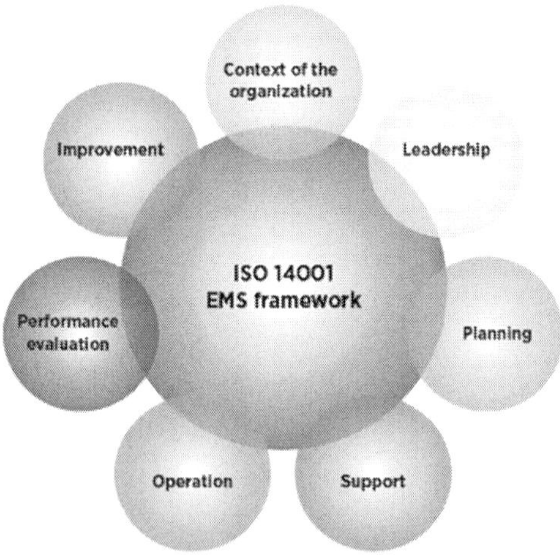

Figure 6.10: ISO 14001 environmental management systems (EMS) framework.

Source: https://asq.org/quality-resources/iso-14001.

6.10. ISO 14004 GENERAL GUIDELINES ON PRINCIPLES SYSTEMS AND SUPPORT TECHNIQUES

ISO 14004 has 10 sections that mirror the corresponding sections in ISO 14001 and for practical sickness, we can look at them being very close relatives of each other they were written concurrently and published before organizations began establishing the IMS to conform to ISO requirements. Currently it is undergoing some substantial change but its general characteristics are that it points to more practical information based on user experience over the last half decade. The major challenges in this section were remaining consistent with the ideas represented in ISO 14001 that are fluid in nature. The most the most important thing to notice is that it is a guidance document going beyond the concepts of ISO 14001 but presents the same basic picture on how to design and implement a successfully functioning environmental management system. Because some of the information in ISO 14004 is a little repetitive and refers back to ISO 14001 the structure of the clauses is given in brief detail below:

- **Clause 1:** Scope defining standards they provide guidance in establishing implementing maintaining and improving EMS in conjunction with other management system.

- **Clause 2:** Normative references however, none are cited.
- **Clause 3:** Terms and definitions that define some terms that are used in the overall document.
- **Clause 4:** EMS that covers clause 4.1 through 4.6.
- **Clause 4.1:** Discusses the general PMS model top management and leadership commitment defining the scope to operate when initializing the EMS. A practical help section is provided as well.
- **Clause 4.2:** Talks about how to design an environmental policy.
- **Clause 4.3:** Discusses planning and in relation to the planning requirements of ISO 14001 with practical help and general guidance.
- **Clause 4.4:** Implementation and operation.
- **Clause 4.5:** Checking expounds on the previous expounds on the previous topic and provide more guidance on implementing and checking requirements of ISO 14001.
- **Clause 4.6:** Management review explains the necessity and details on how to conduct a management review.

Relation to the 1996 and 2004 versions the ISO 14004 designed published in 2016 difference in that it offers way more tangible and practical examples on how to use the given models to create a sufficiently successful EMS. It comes with the underpinning of principles presented in ISO 14001 and explains how to apply them within their respective EMS regardless if one was existent before or not. It can be stated that this is particularly important and useful for new users.

6.11. ISO 14006 A GUIDE FOR ECO DESIGN

ISO 14006 was first published in July 2011 as a guideline for incorporating eco-design into EMS. ISO 14006 is a practical tool used by organizations for the systematic approach to I meant an Eco design as part of the environmental management system. An Eco design is both a principle and an approach integrating Environmental Protection activities with a product's lifecycle. Traditionally organizations would consider products and the impacts only in the manufacturing stage and would neglect the concept of a life cycle which is a cornerstone of eco-design. Essentially eco design is meant to monitor environmental impacts a product would have from production stage all the way to the end of its life cycle and it is applicable to all organizations. The successive stages of products begin from raw material extraction and supply, to manufacturing, product distribution use by customer and finally end-of-life where recovery or recycling of the product occurs.

This section is divided into five chapters. The subdivision is listed as follows:

- Scope;
- Normative references;
- Terms and definitions;
- Role of top management in eco design Guidelines for incorporating eco-design into an EMS;
- Eco-design activities in product design and development.

In general, the topic explains the processes of improving resource and process efficiency, minimizing liability and reduction of costs at all stages of design in a product life cycle. It carries on the emphasis of top management being the fundamental contributor in affecting change without leaving out everyone's involvement in the company. The development of the development of an eco design project is listed in stages from:

- Specifying the product functions;
- Defining its environmental parameters;
- Identifying Environmental Protection strategies;
- Develop environmental objectives and targets;
- Establishing product specification;
- Develop technical solutions.

Other considerations include quality management, human resource, legal, health, and safety requirements all of which can be applicable to the development of an integrated eco design. The idea here was the idea here was to impact change through internal structures by evaluating and implementing innovations in all processes of product design. The costs and problems that come along with implementing an Eco design are essentially negligible and it helps the overall Environmental management system by going beyond just evaluating the environmental output performance (Morris, 2004).

6.12. ISO 14015 ENVIRONMENTAL ASSESSMENTS OF SITES AND ORGANIZATION (EASO)

The international standards provide guidance on how to perform environmental assessments (EAs) on sites and organizations through a specific systematic process of identifying environmental issues and aspects and correlating it to their business consequences. The section contains 5 clauses (Scope, terms, and definitions, rules, and responsibilities, assessment process and reporting) with the fourth showing a guideline on the actual process in which EAs of sites are to be conducted by an organization. Conducting an EA on site and organizations is given in Figure 6.11.

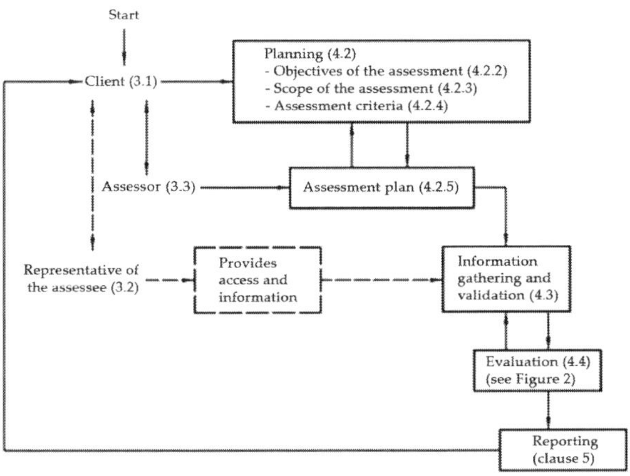

Figure 6.11: Process for conducting an environmental assessment of sites and organizations.

Source: https://www.iso.org/obp/ui/#iso:std:iso:14015:ed-1:v1:en.

6.13. ISO 14020 ENVIRONMENTAL LABELS AND DECLARATIONS, GENERAL PRINCIPLES

Green stickers on consumer products have been present for over 50 years and the main proponents of the idea have been energy and fuel efficiency. Stickers first started appearing in the United States and Canada as legislative requirements developed by government agencies on all sorts of consumer products. It has now moved beyond the traditional borders proving that consumers really do care about the environmental ramifications of the products they buy.

Eco-labeling is a new phenomenon with growing global concerns on Environmental management in government, business, and in the general public. Green stickers and eco labels are labeling systems for products that are a form of sustainability measurements intended especially for consumers to take environmental concerns into account. ISO 1422 1425 is a series of Environmental label declarations where the international standard proposes three categories type 1 in ISO 14024, type II in ISO 14021 and type III in ISO 14025. Generally, they generally are meant to Foster more economically friendly production and consumption within and outside of the organization (Whitelaw, 2012) (Figure 6.12).

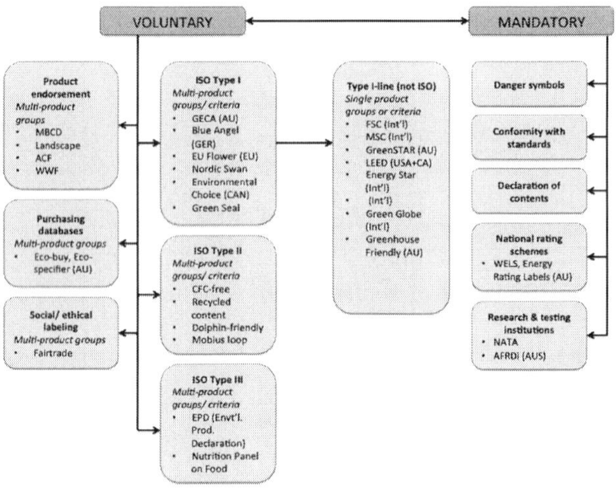

Figure 6.12: Classification of eco-labels.

Source: https://en.wikipedia.org/wiki/Ecolabel.

6.14. ISO 14031 ENVIRONMENTAL PERFORMANCE EVALUATIONS

This section of the international standards provides practical tools techniques and processes for conducting an environmental performance evaluation as part of the overall environmental management strategy. These EPIs make sure that governments document the step-by-step process that they are taking to prevent pollution, correct existing problems and meet legal environmental requirements (Whitelaw, 2012).

The chapter is divided into 4 clauses; the first three in relation to introductory matters and the last one explain the details of environmental performance evaluation. The general overview dictates that a successful EMS should have key performance indicators used to measure the organization's success, and the three kinds have already been explained above (Operation Performance Indicators, OPIs, MPIs, and ECIs. Together with the plan-do-check-act management model it is very straightforward and achievable to meet your environmental performance objectives as an organization.

To build on what was discussed earlier measurements of ECIs, MPI's, and OPIs together with the plan-do-check-act management model does not mean the process is complete. Analyzing data is important and existing quantification methods are available. For example, it is very straightforward to calculate your rate of airborne pollutants produced in your organization. Sometimes it may be useful to sometimes it may be useful to have an independent external monitoring system free of biases that could affect your data. The reason I saw 14031 can be especially youthful is it risk based perception of pinpointing issues, for example, having more than one environmental performance indicator (Whitelaw, 2012).

ISO 14031 continues to enforce the general ISO 14000 rules of management review making sure the best possible information reaches decision makers at the top of the organization and highlights the importance of communication to facilitate change.

6.15. ISO 14040 LIFE CYCLE ASSESSMENTS

Life cycle assessment is one of the new analytical tools provided in the ISO 14000 series as a way to critically analyze your products through all stages of its life cycle allowing a better insight into all possible environmental impacts. Ecological well-being, resource efficiency, and human health are the three areas of study under the life cycle assessment.

The chapter includes the three typical introductory clauses, scope, normative references, and terms and definitions sections. The subsequent three chapters are on general descriptions of life cycles and its principles, methodological framework and reporting (Whitelaw, 2012).

An easy example to offer would be in the car production industry. The life cycle begins with the life cycle begins with gathering raw materials that the car will need during metals from the ground where it is mined and chemicals what battery acid and plastic from plants the sun that is to be process to make Windows and rubber from rubber trees. One can already see the large amount of resources required and after gathering and processing needs raw materials they must be transported to the assembly plant. Next, the car is put together by other people and robots and these robots also use electricity that has an impact on the environment. Once the vehicle has been built, it is then taken to the dealership using more fuel than energy and bought by a happy owner who pumps even more fuel and greenhouse gases (GHG) into the atmosphere. The car requires to be continuously maintained and after years, maybe decades it will be scrapped because it is too old or too broken to be fixed. Life cycle analysis ensures that the most efficient steps are taken in each step here to make sure there is less and less impact on the environment (Whitelaw, 2012) (Figure 6.13).

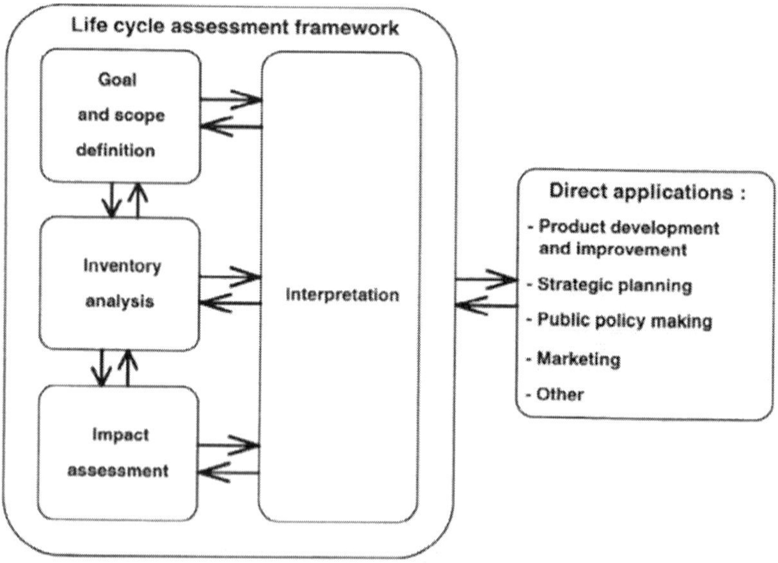

Figure 6.13: Phases and applications of a life-cycle assessment (LCA).

Source: https://www.researchgate.net/figure/Phases-of-an-LCA-7_fig1_339130211.

6.16. PRACTICAL STEPS FOR ISO 14001 CERTIFICATION

There are two types of there are two types of certification under the ISO 14001:

- Certification of a company's environmental management system based on ISO 14001 requirements; and

- Individual certificate to be able to audit on ISO 14001.

For an organization to become certified it must implement an environmental management system based on ISO 14001 requirements and then hire a certified, recognized body to audit on the same grounds. The first step is to define environmental objectives and environmental policy. This begins from the managerial level identifying legal requirements for the EMS.

Other mandatory processes and procedures 4 operations within the organizations should be established and included ensuring they are well documented. The documentation can be done internally by a company or with external consultant help. Once the environmental management has been setup, it needs to run for a period of time to enable data collection and finally auditing and review, before becoming fully certified.

With all the documentation in place and the processes have been implemented, your company needs to follow three steps to ensure successful certification:

- **Internal Audit:** An internal audit needs to happen to ensure proper functioning of EMS processes. This also confirms compliance and mix record keeping a natural part of the system.
- **Management Review:** Formal management review is an integral part of the evaluation process and ensures designation of resources and appropriate decision making.
- **Corrective Actions:** To reinforce the policy of continuous improvement corrective actions need to follow the internal audit and management evaluation.

From here certification may begin and is divided into two stages:

- **Stage 1:** Documentation Review: Internal auditors are to check and ensure documentation meets ISO 14001 requirements; and
- **Stage 2:** Main Audit: Certificate body auditors stepping to monitor the activities and compliance with ISO 14001 standards by reviewing documents records and practices.

6.17. ANALYSIS OF THE ISO 14000

ISO 14000 is one of the only systems that tackle economic, social, and environmental aspects in a way that is objective, proactive, and efficient.

This is known as the triple bottom line. Some of the reasons the ISO 14001 so important are listed below:

- Mitigation of risk by assessment.
- Leading by example by ensuring the organization's top management or fundamentally involved in EMS processes.
- Tax incentives offered by governments both local and state after following ISO standards.
- Employee motivation is boosted when they work in a place that devotes efforts to causes bigger than themselves.
- Brand imaging through fostering environmental policies is long-lasting.
- Proper use of resources is encouraged knowing that natural resources like fossil fuels are finite. Winning companies will encourage the use of alternative resources for their products and services.
- Sustainability is a good stimulus for companies to remain viable and competitive (Figure 6.14).

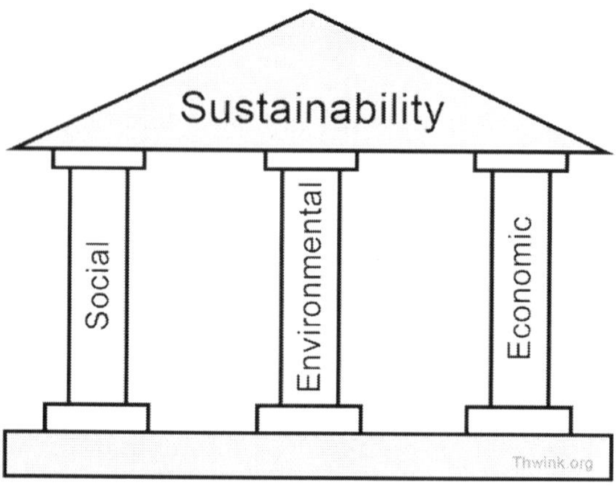

Figure 6.14: Pillars of a sustainable organization.

Source: https://www.process.st/iso-14000/.

The benefits of ISO 14000 compliance includes improvement in:

- Environmental performance;
- Internal management;
- Company image;

- Public relations;
- Government relations;
- Profits;
- Trade;
- External inspections;
- Corporate social responsibility;
- Resource allocation and organization stakeholder satisfaction;
- Overhead costs;
- Insurance cost;
- Cost of production;
- Waste reduction.

According to the World Bank an estimated 50% of pollution created by an unregulated plant can be prevented with minimal investment using improvements such as those outlined in the ISO 14000 family.

Once the ISO 14000 is in place, building upon it is an easy task like employing the ISO 9000 which is a separate standard for quality management systems and continuous improvement can be used in conjunction. When combining standards from both families, obvious care must be taken especially when dealing with conformance and expectations.

Critics of the ISO 14000 complain about the added costs and extensive documentation bring an element of bureaucracy that is unneeded also the potential consequences of uncovering potentially harmful practices can hurt the company's capital insurance and their already existing permits may be compromised (Whitelaw, 2012).

The intent of an international standard is to avoid having multiple standards that may differ, however, the creation of ISO 14000 has not stopped the multiple registration schemes which are presented the world over.

Other issues such as third party certificates don't sit well with many companies especially in the United States where the assessment does not separate the meeting and surpassing of standards with just barely making it. There is a question of who certifies a third-party as auditors can become a big business and manipulated for personal calls.

The last cost with ISO standards is that they do not reflect the diversity of businesses across the world and by extent their specific environmental issues. Small businesses, for example, will have a hard time dealing with the expenses and certification while at the same time knowing that they have to get it in order to compete in the international market. Big businesses on the other hand that operate on the multinational platform complain that the standard doesn't relate to the reality of how they function. The general consensus is there a proper analysis must be done by organizations before entering the ISO 14000 because many already have the EMS in place and may not need to replace them (Whitelaw, 2012).

7

Environmental Ethics

CONTENTS

7.1. INTRODUCTION

Environmental moral values are a division of ethics that examines the connection between the human race and the surroundings and how ethics play a major part in all this. As a consequence, it is very vital that each individual respects and gives honor to this and fully implements the morality and the moral code when they are dealing with these particular mortals. Environmental ethics is also an established area of the practical beliefs which renovates the essential kinds of the argumentation that may be made for shielding the natural units and the sustainable implementation of all the natural assets. The major challenging paradigms are anthropocentrism, physiocentrism also referred to as the ecocentrism, and ethnocentrism. Environmental ethics exercises the manipulation of a huge range of these regulations which consists of the environmental regulation, ecological sociology, the ecotheology, the ecological finances, the biology, and the environmental topography. The academic area of the ecological ethics developed up in the retort to the works of Carson Rachel and Bookchin Murray and the procedures like the initial Earth Day in the year 1970, when the environmentalists argued the philosophers to think about the philosophical features of the environmental troubles. Two articles that were published in Science had a very vital impact to White's Lynn. The chronological Roots of the Ecological catastrophe in March of the year 1967 and Hardin's Garrett. The catastrophe of the Commons in December of the year 1968. Furthermore, the other influential was Hardin's Garett and his afterward essay was known as the exploring latest Ethics for continued existence, as well as the thesis by Leopold Aldo in his A Sand Region Almanac, that was known as The Earth Ethic, in which the Leopold unambiguously argued that the pedigree of the environmental crisis were the philosophical in the year 1949 (Boylan, 2013).

7.2. ANTHROPOCENTRISM

Anthropocentrism is the point that humans are the majority important or the critical aspect in any particular situation; that the individual race must constantly be its own initial concern. The cynics of anthropocentrism disagree that the Western institution biases the Homo sapiens when they are bearing in mind the environmental moral codes of circumstances and that individuals evaluate their surroundings or the other organisms in conditions of the utility for them. Majority of the people argue that all surroundings studies can be acquired into reporting the assessment of the inherent value of the non-human creators. In verity, founded on its very theory, the philosophical editorial has searched recently the prospect of the humans' enthusiastic extermination as a gesture on the road to other individuals. The authors pass on the thought as a consideration experiment that would not be unspoken as an entitled for the action. Spinoza Baruch logic that the humans were to search for equipment objectively, they could realize that the whole thing in the universe has a unique importance. Similarly, it is achievable that a creature-centered, anthropocentric, or, androcentric ethic cannot be an accurate representation of reality, and there would be a bigger picture that individuals can or cannot be capable to comprehend from the human standpoint (Boylan, 2013).

Vardy Peter distinguished stuck between the two kinds of anthropocentrism. A physically powerful anthropocentric ethic squabbles that the human race is at the middle of authenticity and it is correct for them to be like that. The feeble anthropocentrism, conversely, argues that actuality may only be taken to mean from a human position of view, hence the individuals will have to be at the middle of the veracity as they perceive it well.

Another position of scrutiny has been fashioned by Norton Bryan, who has lately developed into one of the indispensable actors of the ecological ethics by the launching of the environmental expediency, and now it's one of its leading developments. Environmental expediency rebuffs to take a posture in disputes amid the protectors of the anthropocentric and the nonanthropocentric ethics. In its place, Bryan differentiates between the physically powerful anthropocentrism and the feeble or- the comprehensive-anthropocentrism and creates a row that the previous must undervalue the diversity of the instrumental morals humans can be obtained from the expected world (Keller, 2010).

A current view narrates anthropocentrism to the prospect of life. The biotic moral values are founded on the human distinctiveness as part of the genetic material or the protein organic existence whose effectual purpose is the self-dissemination. This entails a human reason to protect and disseminate life. The human races are in the middle since only they may secure existence beyond the period of the Sun, probably for the trillions of the eons. The biotic moral code values existence itself, as the personified in the biological organizations and the processes. The individual is the special since they may be locked and the prospect of existence on the cosmological balance. In meticulousness, human's beings may continue conscious life that takes pleasure in its existence, accumulating further the enthusiasm to propagate the existence. The human being can protect the expectations of its existence, and these expectations can provide the human continuation a huge purpose (Keller, 2010).

7.3. HUMANIST THEORIES

Singer Peter advocated the maintenance of the world heritage position, the unspoilt sections of the world that obtain an insufficiency worth as they lessen over occasion. Their conservation is an inheritance for the future cohort as they have already been inherited from the human's ancestors and would be conceded down to the future cohort so they may have the chance to make their minds up whether to take pleasure in the unspoilt countryside or the exclusively urban scenery. A good paradigm of a world legacy site could be the sultry tropical forest, a very expert ecosystem that has taken centuries to develop. Clearing the tropical rain forest for farmland usually fails due to the soil circumstances, and on one occasion disturbed, may obtain thousands of years to the redevelopment.

7.4. MORAL THEORIES

7.4.1. Momentous Theories

Consequentialism is the category of the normative moral theories holding that the penalty of one's behavior is the decisive foundation for any verdict about the appropriateness or the injustice of that manner. Therefore, from a consequentiality position of the observation, and ethically exact performance or the failure to notice from the substitute is one that will produce a good termination, or the outcome.

Consequentialism is the initial non-prescriptive; significance the ethical worth of a deed is strong-minded by its possible consequence, not only by whether it permits a set of the written proclamations or the laws. One instance would involve lying beneath the threat of the administration's punishment to put away a blameless person's existence, even if it's against the law to lie under pledge (Keller, 2010).

Consequentialism is more often than not contrasted with the deontological moral principles or the deontology, in that particular deontology, in which the set of laws and the moral obligation are in the central, which obtains the aptness or the injustice of one's behavior from the nature of the performance itself somewhat than the ending of the conduct. It is also compared with the virtue ethics, that is centered on the personality of the mediator relatively than on the personality or the consequences of the performs or the omission itself, and the pragmatic ethics that cares for the morality reminiscent of science the advancing collectively over the itinerary of the various lifetimes, also as to of any moral decisive factor is matter for revision.

Consequentiality hypothesis differs in the customs they can only explain the ethical goods. A number of people bicker that consequentiality and the deontological conjecture are not inevitably mutually elite. For instance, T. Scanlon enhances the suggestion that human privileges, which are frequently considered as deontological impression, may only be defensible with the orientation to the consequences of comprising those rights. Correspondingly, Nozick Robert disagreed for an assumption that is frequently consequentiality, but integrates inviolable the side-limitations which limit the sort of the actions causes are acceptable to do so. Parfit Derek squabbled that in performance, when implicit properly, statute consequentialism, the Kantian deontology, and the contractualism could all finish up prescribing the equivalent (Keller, 2010).

7.4.2. Deontological Hypothesis

In the proper philosophy, the deontological morals or the deontology is the normative moral hypothesis that the ethics of an achievement should be founded on whether that specific achievement itself is the precise or erroneous under a sequence of the rules, to a certain extent than founded on the consequences of the achievement. It is occasionally described as the responsibility-, requirement- or the rule-foundation ethics. Deontological beliefs are commonly disparity to the consequentialism, the virtue ethics, and the sensible ethics. In these specific expressions, the accomplishment is the extra considerable than the consequences (Kernohan, 2002).

The expression deontological was initially implemented to illustrate the current, dedicated explanation by C. Broad in his volume in 1930, the five sorts of Ethical hypothesis. The older convention of the idiom goes way backside to the Bentham Jeremy, who invented it before the year 1816 as a synonym of the Dicastic or the Censorial beliefs that are the ethics that are established on the judgment. The extra broad intellect of the term is maintained in the French, in particular in the idiom ethical regulations, in the circumstance of the professional moral codes.

Depending on the classification of the deontological moral principles under the deliberation, that the moral obligation may occur from an exterior or the interior source, like the set of regulations that are intrinsic to the universe decent naturalism, the religious decree, or a group of the personal or the enriching values several of which can be in an argument with the personal requirements. A good deal of the explanations on the deontological moral values is established in the ancient Tamil prose Thirukural that is whispered to be printed amid the 300 BCE and the fifth century (Kernohan, 2002).

7.5. ATTITUDES

7.5.1. Utilitarianism

In synopsis, Bentham Jeremy affirms that people are obsessed by their benefits and their doubts, but their welfare takes preference over their qualms, and their welfare are conceded out in the agreement with how the community views the penalty that strength is caught up with their interests. Contentment on this version is definite as the maximization of the contentment and the minimization of the pain. It may be squabble that the subsistence of the phenomenal realization and the quail is compulsory for the knowledge of pleasure or the pain in having a principled significance. In history, the self-indulgent utilitarianism is the definitive instance of a consequentialist ethical hypothesis. This appearance of utilitarianism grasps that what is important is the cumulative happiness; the contentment of everybody and not the contentment of a few of the particular individuals. John Mill Stuart, in his description of the riotous utilitarianism, the projected hierarchy of the delight, gist that the detection of certain classes of delight is more exceedingly cherished than the detection of other specific pleasures. On the further supply, some fashionable utilitarians, like Singer Peter, are disturbed with making the most out of the contentment of the favorite, therefore penchant utilitarianism. Other existing forms of the utilitarianism represent the types of consequentialism that are delineating below:

- **Imperative Utilitarianism:** In a broad sense, the consequentialist hypothesis focuses on the actions. Nevertheless, this must not be the folder. The imperative consequentialism is an assumption that is occasionally seen as an effort to reconcile the deontology and the consequentialism and in most cases; this is confirmed as a condemnation of the rule consequentialism. Like the deontology, the rule consequentialism grasps that the ethical behavior entails the following of specific definite rules. Conversely, the rule consequentialism decides the rules established on the penalty that the varieties of those regulations have. The imperative consequentialism subsists in the kinds of rule utilitarianism and the imperative egoism.

A variety of theorists are divided as to whether the regulations are simply the determinant of the moral activities or not. For instance, Nozick Robert apprehended that certain groups of minimal regulations, which he referred to as the side-limit, are essential to ensure the appropriate procedures. There are a number of disparities as to how the absolute of these ethical regulations are. Accordingly, while Roberts's side-limitations are absolute boundaries on the performance, Amartya Sen suggests a hypothesis that acknowledges the importance of confident rules, but these specific rules are not inclusive. This is why; they can be desecrated if the stern observance to the imperative would escort to a lot more of the disagreeable consequences (Keller, 2010).

A line of the major common protestation to the rule of consequentialism is that it's rambling, because it is founded on the consequentialist attitude that what we could be disturbed with is the maximizing the excellent, but subsequently it advises us not to proceed to maximize the best, but to pursue the rules yet in cases where we recognize that the flouting of the rule would manufacture better consequences.

- **Motive Consequentialism:** Another of the consequentialist description is the motive consequentialism which gazes at whether the position of the associations that the outcome from the purpose to desire an action which is better or at smallest amount as better as all of the substitute state of the affairs that could have resulted from the substitute actions. This description gives the importance to the purpose of a performance and connects it to its penalty. A performance can as a result not be erroneous if the verdict to proceed was established on the exact motive. A potential conjecture is that individual which cannot be responsible for the mistaken verdict if the enthusiasm was to do better (Keller, 2010).
- **Pessimistic Consequentialism:** Majority of the consequentialist hypotheses focus on promoting various sorts of excellent consequences. Nonetheless, the negative utilitarianism put down a consequentialist hypothesis that centered solely on the minimizing of the terrible consequences.

One of the majority differences among these two advances is the agent's accountability. Positive consequentialism strains that we convey about the good condition of the affairs, whereas the negative consequentialism necessitates that we should avoid the terrible ones. Stronger descriptions of unhelpful consequentialism will need active involvement to avoid bad and the revolutionize existing impairment. In the weaker sides, uncomplicated forbearance from performance tending to damage others is adequate. An instance of this is the Slippery incline Argument that supports others to stay away from a specified performance on the foundation that it can ultimately direct to the unwanted consequences (Keller, 2010).

Often pessimistic consequentialist hypotheses assert that the reduction anguish is more significant than the increasing contentment. Popper Karl, for instance, maintained from the ethical point of inspection, pain should be overshadowed by the pleasure. While Karl is not a consequentialist per se, this is given as the classic declaration of the negative utilitarianism. When bearing in mind a theory of impartiality, the negative consequentiality can implement a statewide or the global-accomplishment principle and the lessening of distress for the underprivileged which is more precious than the increased delight for the prosperous or the luxurious.

- **Two-Point Consequentialism:** The two-point approach consists of the engaging in the serious reasoning and the allowing for all the potential ramifications of the individuals' actions before creating an ethical verdict, but reverting to the usually reliable ethical rules when the individual is not in a place to stand at the rear and scrutinize the dilemma as a complete. In the performance, this compares the remainder to the rule consequentialism when the lone may only be the reason on an instinctive level, and to perform consequentialism when in a point to stand at the rear and reason on a graver level.

This point may be explained as an understanding between the performance consequentialism that the principles of an achievement is resolute by that trial effects and the regulation consequentialism in which the good behavior is consequent from the subsequent rules that direct to the positive endings (Keller, 2010).

- **Principled Egoism:** The principled egoism can be implicit as the consequentialist hypotheses according to the particular consequences for the lone agent are given to the material more than any extra consequence. Hence, egoism will set down the actions that can be advantageous, the disadvantageous, or even the disinterested to the interests of others. Several as Henry Sidgwick, disagree that confident degrees of egoism endorse the general benefit of the world for two motives: since the individuals recognize how to delight themselves paramount and for that reason, if everybody were a severe altruist then the common welfare could inevitably be reduced.

- **Principled Altruism:** The ethical altruism can be observed as a consequentialist ethic that recommends that an individual take measures that have the most excellent consequences for everybody apart from for himself. This was supported by Comte Auguste, who invented the word altruism, and whose moral principles can be added up in the expression Live for others.

7.6. ISSUES

7.6.1. Action Management

The most important characteristic of the majority of normative ethical theories like consequentialism is the aptitude to create practical ethical decisions. In the very smallest amount, any ethical theory requires to describe the point of view from which the integrity of the penalty is to be resolved. What is principally at chance here is the accountability of the agent.

7.6.2. The Perfect Observer

The main common method among the consequentiality, predominantly those dedicated to the altruistic or the selfless account of the consequentialism, is to apply a model, the neutral spectator from which the ethical ruling which may be made. Rawls John, a detractor of the utilitarianism, quarrels that the utilitarianism is ordinary with other kinds of the consequentialism, which relies on the standpoint of the particular an ideal onlooker. The meticulous distinctiveness of this perfect observer may differ from the omniscient viewer, who could clutch all the penalty of any performance, to a preferably informed witness, who identifies as much as would be practically anticipated, but not essentially all the conditions or the entire probable consequences. The consequentialist speculation that approves this pattern holds that precise action is the performance that will take regarding the best outcomes from this ultimate observer's standpoint (Keller, 2010).

7.7. ENVIRONMENTAL ETHICS IN SPIRITUALITY

7.7.1. Transcendentalism

Transcendentalism is an idealistic movement that was created in the delayed years of the 1820s and the 1830s in the eastern unified countries. It occurred as a response, to remonstration against the broad state of intellectualism and the spirituality at the occasion. The principle of the Unitarian cathedral as was skilled at the Harvard Divinity institution was directly related.

Transcendentalism surfaced from the English and the German impracticality, the Biblical condemnation of Johann Herder Gottfried and Schleiermacher Friedrich, and the disbelief of Hume David and the transcendental idea of Kant Immanuel and the German impracticality. Miller and Versluis observed Swedenborg Emanuel as a persistent influence on transcendentalism. It was as well powerfully predisposed by the Hindu transcript on the philosophy of the brainpower and the spirituality, in particular the Upanishads (Boylan, 2013).

The heart idea of transcendentalism is in the intrinsic goodness of the natives and the natural history. Enthusiast considers that the world and its establishments have tarnished the clarity of the individual; also, they have assurance that the natives are at their unsurpassed when actually self-reliant and self-determining.

Transcendentalism highlights subjective insight over the purpose of empiricism. Adherents consider that the persons are proficient of the production completely inventive insights with the tiny attention and the reverence to the precedent masters. Transcendentalism is strongly associated with Unitarianism, the foremost religious society in Boston in the untimely nineteenth century. It happened to increase after the Unitarianism acquired clasp at Harvard institution of higher education, following the determination of Henry Ware the same as the Hollis lecturer of the Divinity in the year 1805 and of John Kirkland Thornton as the leader in the year 1810. Transcendentalism was not a negative response of Unitarianism; fairly, it was created as the unrefined outcome of the Unitarian importance on the free scruples and the worth of the academic reason. These transcendentalists were not comfortable with the moderation, the warmth, and the tranquil rationalism of Unitarianism. As a substitute, they longed for further concentrated spiritual practice. Hence, transcendentalism was not natural as a contradict-association to Unitarianism, although as a comparable association to the very thoughts that were commenced by the Unitarians (Boylan, 2013).

- **Attitudes:** Transcendentalists are the physically powerful supporters in the rule of the entity. It is principally disturbed with the personal choice. Their attitude is closely associated with persons of the Romantics, but they fluctuate by a challenge to cuddle or, at slightest, to not resist the empiricism of the discipline.
- **Individualism:** Transcendentalists accept as true that the world and its organization mainly organized belief and the biased parties that fraudulent the clarity of the entity. They have assurance that these natives are at their greatest when truly they are self-contingent and self-regulating. It is merely from such actual persons that the factual community may be structured. Yet with this essential independence, transcendentalists besides consider that all natives are opening for the Over-spirit. Since the Over-spirit is one, this fuses all the people as one organism. Emerson alludes to this perception in the prologue the American intellectuals deal with, to facilitate there is One Man, currently to all exacting men only moderately, or even during one ability; and that one must acquire the entire world to discover the entire man. Such an idyllic is in the concord with the Transcendentalist individualism, as all the people are authorized to observe surrounded by him or her section of the marvelous Over-spirit (Boylan, 2013).
- **Authority on Other Schedules:** Transcendentalism is, in various facets, the first prominent American rational movement. It encompasses subsequent creations of the American thinkers, also as well as several literary schedules. Transcendentalism prejudiced the growing association of the cerebral Sciences of the mean-19th century, which would afterward become acknowledged as the New Thought progressed. The latest thought regards Emerson as its scholar father. Curtis Emma Hopkins the trainer of all trainers, Holmes Ernest, originator of the spiritual Science, Charles, and the Fillmore Myrtle, originator of the harmony, and Cramer Malinda and Nona Brooks, the originator of the celestial Science, were all the seriously predisposed by Transcendentalism (Boylan, 2013).

Transcendentalism is partial to Hinduism. Ram Roy Mohan in the year 1772–1833, the originator of the Samaj Brahmo, redundant Hindu mythology, however, also the Christian trinity. He establishes that the Unitarianism arrived adjoining to the actual Christianity, and they had a physically powerful consideration for the Unitarians, who were strongly associated with the Transcendentalists. Ram Roy Mohan established a missionary commission in Calcutta, also in the year 1828 they requested for sustenance for the follower's actions from the American Unitarians. By the year 1829, Mohan had discarded the Unitarian commission, however, after Mohan's death, the Samaj reserved close fasten to the Unitarian cathedral, who struggled towards a balanced reliance, the common reorganization, and the amalgamation of these two in a transformed religion. Its mysticism was referred to as the neo-Vedanta by the Christian observers, which has been extremely significant in the current popular understanding of Hinduism, however, also of the recent western theology, which re-introduced the Unitarian controls in the cover of the apparently age-mature Neo-Vedanta.

7.7.2. Deep Ecology

Deep ecosystem is a biological and the environmental attitude promoting the intrinsic significance of the living creatures regardless of their influential utility to the human requirements, plus a reorganization of the modern human humanity in agreement with such particular thoughts.

Deep environmental science quarrels that natural humanity is a delicate balance of multipart inter-associations in which the continuation of the organisms is reliant on the continuation of others inside the ecosystems. Human intrusion with or the demolition of the accepted world fronts a threat thus not only to the individuals but to all the organisms that make up the natural arrangement (Keller, 2010).

Deep ecosystem is the core belief is the principle that the living atmosphere as a whole would be esteemed and observed as having confident basic descent and the legal privileges to live and to prosper, independent of its helpful benefits for human implementation. Deep natural science is often surrounded in conditions of the scheme of a great deal broader sociality; it acknowledges the diverse groups of life in the globe that are collected not only via the biotic features but as well as, where appropriate, through the moral affairs, to facilitate it, the worth of other creatures as more than immediately resources. It explains itself as being deep since it considers itself as appearing more extremely into the genuine actuality of humanity's relationship with the normal world incoming at the philosophically more insightful assumptions than that of the established view of the ecosystem as the branch of ecology. The association does not pledge to the anthropocentric conservatism that is apprehensive with the maintenance of the atmosphere only for the development by and for the human idea, ever since the deep natural science that is stranded in a rather different group of the philosophical hypothesis. The deep ecosystem takes a further holistic observation of the globe human creatures live in and search for the appropriate to life and the sympathetic that the separate sections of the ecology which comprises the humans all function as a complete. This attitude provides an establishment for the ecological, the environmental science, and the green arrangements and has promoted a new structure of the environmental moral codes that are advocating the rough country preservation, and the human inhabitant's control, and the simple livelihood (Keller, 2010).

- **Growth of Deep Ecosystem:** The expression of deep ecosystem was invented by the Norwegian theorist Arne Næss, in 1973. Arne discarded the suggestion that beings may be categorized according to their virtual value. For instance, the ruling on whether the creature has a perpetual soul, whether it implements the reason or even whether it has realization or the certainly of higher realization have all subsisted to justify the position of the human creature as the advancement to the other natural world. Arne maintains that from an environmental point of observation the correction of all kinds of existence to live is a worldwide right which would be enumerated. No solitary species of the living mortal has extra of this exacting right to exist and spread out than any other genus. The environmental problems that are expressed by the globe today are moderately due to the defeat of the traditional awareness, the ethics, and the moral code of activities that commemorate the intrinsic assessment and the holiness of the natural globe that provide the preservation of the scenery's prime significance. In the same way, the hypothesis of the human pre-eminence to other existence structures, as if we were approved royalty ranks over natural history, the suggestion that Nature is principally here to provide human will and the idea, obtains a radical appraisal in the deep ecosystem. Deep ecosystems created a reaction to anthropocentric analysis and the several dissimilar participants who participated in an important chronological role in its augmentation. Important amongst them was Joseph Meeker, who in the year 1973 narrated to Sessions George regarding Arne Næss, whom Meeker knew in my judgment. As Warwick related, an individual of the stuff that initially attracted George about Arne was Næss's tough curiosity in, and ground-breaking advance to, the efforts of Spinoza. George narrates that he had himself reached at Spinoza as the response to the procedure of teaching narration of the attitude by concerning in the year 1972 and separately of being in touch with Arne. George consequently wrote to Arne at this time, commencing a lifelong involvement. Meeker's in the year 1972 and 1997 in the volume The Pleasantry of Survival: learning in the literary ecosystem appeared through the effort of scholars in quest of an ecological ethic. That volume stands for Joseph's founding work in mythical environmental science and the ecocriticism that reveals the relationship connecting the mythical arts and the technical ecology, in particular humankind's deliberation of comedy and catastrophe. It prompts the readers that the adaptive manners in comedy to encourage survival, where catastrophe estranges from extra life shapes. This hypothesis rests on Joseph's study of proportional literature, his effort with the environmentalist Lorenz Konrad, and his efforts as an area natural scientist in the National Square Service in Alaska, in Oregon, and California.

- **Environmental Tutoring:** Ecosystem in the willowy sense which is known to the natural discipline of the natural balance. Nevertheless, the environmental paradigms and the main beliefs are being created and applied in approximately all the disciplines, and these particular paradigms have to perform with the technique we advance accepting the relationships and the inter-associations within and among the living creatures which provide to each its individual place and the characteristics. The human ecosystem, for example, must definitely take a description of the position of our prejudiced lives and the spiritual requirements, also the biological ones, in conditions of their ecological things. The environmental science in this logic may not be reductionist responsibility, but a group toward another complete or the holistic image and the understanding of the world's procedures. Deep conservationism seeks to observe into all the levels of continuation and may be observed as being radical by several; for all of them other anthropocentric observation is appropriate since it puts the human beings at the middle. Studying how to live in synchronization with our environments is advantageous because preventing the global destruction crisis and attaining actual environmental sustainability will entail rethinking the principles as a culture. To facilitate the way teaching appears to be the most excellent way to begin. Sustainability teaching aims to assist learners to appreciate their interconnectedness amid all existences, to develop into creative crises solvers and the active society, and to connect personally and mentally in determining our common expectations. Experimental knowledge and significant pedagogy are in the middle to provide chances for the learners to connect in the sustainability scholarship. The surroundings broadly definite, vestiges somewhat abandoned within the expansion studies, although a substantial augment in the donations to the area more than the fifteen years ever since 2000.

Scholars and the postgraduate programs with several prominent exceptions frequently add on the ecological issues as extraordinary lessons or the modules, also their relics a tendency for the ones who are stranded in the substance and the discursive great efforts that describe the control to believe the surroundings as the exotic individual interest, a trouble marks itself in cultures that have the relaxation to concerned about the usual world. Expansion of a modern teaching model promoting nationalism and civic accountability, the vigorous social situation and the healthy existence is closely connected to the development of the ecological responsibility in the younger age bracket. An expansion of the environmentally accountable character in the character is of particular significance for the former students of educational organization. Environmental teaching would be incorporated in the different prospectus in majority of the fields: schooling for sustainable expansion in the circumstance of ecopedagogy (Keller, 2010).

7.7.3. Judeo-Christian Custom

The thought that an ordinary Judeo-Christian principle or the Judeo-Christian moral underpins the American affairs of state, the regulation, and the morals has become a section of the American general belief ever since the year 1940s. In the current years, the expression has been related with the American conservatism; however, the perception despite the fact that not always the accurate slogan has normally featured in the oratory of leaders transversely the political range, consisting of that of Franklin Roosevelt and Lyndon Johnson (Kernohan, 2002).

- **Criticism and Answers:** Several theologians advise against the gullible implementation of the Judeo-Christian entirely, disagreeing that it may license disobedience, like the opposition to the material humanism with regard to the present Jewish, the Catholic, or the Christian ethnicity, which consist the tolerant strains of the various faiths, for example, the modified Judaism and the freethinking Protestant Christianity.

Two prominent books tackled the relations among current Judaism and the Christianity. Abba Silver Hillel's Where Judaism varies and the Baeck Leo's Judaism and the Christianity were equally motivated by the impulse to elucidate Judaism's uniqueness in a globe where the word Judeo-Christian had covered the critical variations amid the two devotions.

Reacting alongside the blurring of the theological features, Rabbi Berkovits Eliezer recorded that Judaism is Judaism since it discards Christianity, and the Christianity is Christianity since it discards Judaism.

Theologian and the novelist Arthur Cohen, in the legend of the Judeo-Christian custom, had uncertainty in the theological authority of the Judeo-Christian idea and he recommended that it was basically an development of the American government, even with the actuality that Jacob Neusner, in the Jews and the Christians: the legend of a ordinary Tradition, records, that the two beliefs stand for various people discussing about the different effects to the different citizens.

Law lecturer Stephen Feldman, gazing at the era before the 1950, primarily in Europe, observes that the impression of a Judeo-Christian custom as the supersessionism, that he describes as the treacherous Christian system of belief in the slightest from a Jewish perception, also as a fairy tale which insidiously incomprehensible the actual and momentous differences amid the Judaism and the Christianity.

- **Abrahamic Belief:** The supporters of the word Abrahamic belief ever since the subsequent half of the twenty century have projected an inclusivism that broadens the Judeo-Christian notion to consist of the Islam also. The underlying principle for the expression Abrahamic is to facilitate Islam, for example, Judaism and the Christianity, finds its genesis to the stature of Abraham, whom is also considered as the prophet in Islam. Activists of this sunshade term believe it as the investigation of incredible positives in the logic of the spiritual connection amid the Jews, the Christians, and also the Muslims.

- **Australia:** Australian historian Taylor Tony directs out that Australia normally has rented the Judeo-Christian subject from the American traditionalist conversation. Jim Berryman, one more Australian historian, disagrees that from the year 1890s to the current, oratory upholding of the Australia's conventional connection to the Western empire emphasizes the three idea: the center British legacy the Australia's Judeo-Christian trust the system and the balanced ideology of the clarification. These ideas have been articulated frequently on the Australian core-accurate political continuum, and the majority significantly among the unadventurous-leaning observers.

7.7.4. Oriental Religions

The Eastern creeds are the creeds that were invented in the East, the South, and the Southeast Asia and hence have variations with the Western beliefs. This consists of the East Asian beliefs like the Shintoism, the Sindoism, the Taoism and the Confucianism, also the Indian creeds consists of the Hinduism, the Buddhism, the Sikhism and the Jainism also as well as the animistic native creeds (King-Tak, 2009).

This East-West creeds feature, immediately as with the East-West mores division, and the proposition that happens from it, is extensive and not accurate. In addition, the geographical division has less denotation in the current world.

While the majority of the Western onlookers attempt to differentiate between the Eastern attitudes and the beliefs, this is a division that does not subsist in the majority of the Eastern backgrounds.

- **Indian Beliefs:** Beliefs began in the Indian subcontinent that consists of Hinduism, Buddhism, Jainism, and Sikhism. The theologies and the ideas of these religious convictions have had several conceptions in common, for example, the dharma, the karma, the maya and the samsara.
- **Hinduism:** It started off in the Indian subcontinent. In addition, it is well thought-out by several to be the globe's oldest existing major belief. Hinduism consists of a vast body of the scripture,

alienated as the exposed and the remembered, descriptive of the dharma, or the religious existing. Hindus reflect on the Vedas and the Upanishads as creatures among the prime in the influence, significance, and the ancient times. The Bhagavad Gītā, a conversation excerpted from the Mahābhārata, is also known as a summing up of the saintly acquaintance of the Vedas. It is complex to recognize any universal principles or the practice in Hinduism, even though prominent subject matter consists of the Dharma, the Samsara, the Karma, and also the Moksha. Hinduism is occasionally referred to as the polytheistic creed, however, this is a generalization. Hinduism consists of a diverse anthology of schools whose ideas span monotheism, polytheism, pantheism, monism, and even atheism. For example, the Advaita Vedanta institute grasps that there is merely one causal article which is Brahman, which is evident itself into all the diverse existing and the non-existing kinds that we scrutinize, whereas ethnicity like the Vaishnavism and the Shaivism worship the Vishnu and the Shiva in a comparatively more monotheistic intelligence that has grounds by a differentiation amid the parabrahman and the atman. An amount of scholars even believes the Samkhya institute of contemplation to have atheistic leanings.

- **Buddhism:** It is the nontheistic Indian conviction and the ideals. Buddhism was established around the fifth century in India by Siddhartha Gautama, also the Buddha, with the four dignified reality and the Eightfold pathways as its core ethics. However, according to the scriptures, the Four Noble realities were exposed by the Buddha in his initial sermon after acquiring the explanation. The institutions of Buddhism are naturally divided into the Theravada and the Mahayana. In the intellectual spheres, Mahayana is supplementary divided into East Asian and the Tibetan Buddhism. This Buddhism educates that the individual who becomes free-thinking without the teaching is the Buddha. The initial aim of Buddhism is the release of the practitioner from the samsara. The Buddhists grasp all this to be the resolution to the trouble of the anguish.
- **Sikhism:** It is a creed which is created in the Punjab of Northern India. However, it was established on the learning's of the Guru Dev Nanak and the nine individual gurus that were consequent. He usually has an idea to preach the route to explanation and

God in Sultanpur. His observations rejected the long-established worships and the background of the Hindu creed. Liberty from the re-embodiment which is tied to commemoration and the recurrence of one widespread God. God is loose-fitting and concurrent in every kind. Sikhs consider that there is only a sole general God who is the final creator, the sustainer, and the destructive force. The Gurū Granth Sāhib is the core scriptures were anticipated to conserve the hymns and the learning's of the Sikh Gurus and the further saints from the Hindu and the Sufi ethnicity. The sacraments, the religious rituals or the blank worship are well-thought-out to be of little implementation and the Sikhs are from the fasting or the departure on the pilgrimages. The doctrine of Sikhism consists of (1) the sincere living or the earning, (2) the tithing and the giving of the alms (3) the vocalization of God. Sikhism moreover has a physically powerful warrior ritual which happens in resistance to the religious autonomy and the human privileges from an autocratic Moghul profession of India.

- **East Asian Religions:** The collection of the East Asian creed or the attitude, additionally known as the Taoic creeds, that can be well-thought-out as the variety a group surrounded by the world creeds equivalent to the Abrahamic or the Dharmic assembly. The Taoic beliefs have at the lowest 500 million followers in the worldwide.
- **Taoism:** Which is acknowledged as Daoism, includes a mixture of the related devout and the philosophical customs. The tagging of the Taoist factions and the movements is very contentious. Taoist decorum and the principle place the prominence on the three charms of the Tao which are made up of love, restraint, and modesty. Taoist religious studies focus on the set of guidelines of the Wu Wei the non-action, impulsiveness, humanism, relativism, and the bareness. The majority of the conventional Chinese Taoists are the polytheistic. There is a disparity concerning the suitable masterpiece of this pantheon. The accepted Taoism usually presents the Jade monarch as the top deity. The thinker, or the elite, Taoism typically presents the Laozi and the Three unpolluted Ones at the pinnacle of the pantheon. The natural history and the ancestor feelings are widespread in the well-liked Taoism. Nevertheless, this species of the shamanism which

is shunned for an importance on the interior alchemy amid the influential Taoists. The Tao itself is hardly ever an object of reverence; the individuals are treated mostly as the South Asian perception of the atman.

- **Shinto:** This is an animistic folk creed that is experienced in Japan. Shinto accurately means the route of the deity. Majority of the Japanese Shintoists furthermore recognizes themselves as Buddhists. The Japanese unpolluted earth Buddhism which is deeply entwined with the Shinto creed. The Shinto practitioners commonly announce tradition, the relations, the nature, the cleanliness and the formal procedure observation as the core standards. Taoic authority is momentous in their viewpoint about nature and the personality-mastery. The sacrament cleanliness is a core part of the Shinto existence. The holy place has a momentous place in Shintoism, shimmering the animistic adoration of the kami. The folks, or the accepted, the Shinto places an importance on the shamanism, predominantly in the prediction, the spirit ownership, and the faith therapeutic. The faction Shinto is a varied group which consists of the mass-worshippers and the Confucian Shintoists.
- **Confucianism:** It is a difficult coordination of the moral, the common and the political deliberation, significant in the narration of East Asia. It is controversial whether Confucianism is a creed or merely an ethical structure. It is generally associated with legalism but it castoffs the legalism for ritualism. It furthermore endorses the meritocracy as the supreme of the graciousness. Confucianism has an intricate classification governing obligations and the good manners in the relationships. Confucian moral principles focal point on the familial task, the trustworthiness, and the humaneness. Confucianism stands in the Chinese folk acknowledgment of the continuation of the animistic emotional state, the ghouls, and the goddess. It endorses paying them the appropriate esteem, but at a further fundamental rank that encourages evades them. Confucian deliberation is as prominent as the scaffold upon which the syncretic Neo-Confucianism was created.

7.7.5. Animism

Animism is the conviction that the objects, the places, and the individuals all possess a dissimilar spiritual quintessence. The potentially, animism perceives all the stuff which include the animals, the flora, the rocks, the rivers, the conditions structures, the human achievement and possibly even the words as the energetic and the active. Animism is implemented in the anthropology of the creed as an expression for the belief structure of the majority of the native peoples, in particular in distinction to the comparatively more current development of the organized religious conviction (King-Tak, 2009).

Even though each ethnicity has its individual different legends and the rites, animism is believed to be illustrated as the most familiar; the initial filament of the original peoples the sacred or the mystical point of view. The animistic standpoint is so generally held and intrinsic to the majority of the indigenous commoners that they regularly do not yet have a speech in their speeches that communicate to the animist or yet the religion, the phrase is the anthropological assemble.

The principally outstanding to like the ethnolinguistic and the cultural inconsistency, the opinion has been diverging on whether the animism is known as the inherited mode of the occurrence common to the local peoples approximately the humanity or to a packed-fledged creed in its own correctness. The currently established classification of the animism was only created in the late 19th in the year 1871 by Sir Tylor Edward, who twisted it as one of the anthropology's initial impressions, if not the initial.

Animism covers the ideas that all material occurrence have outfit, that there subsist no hard and speedy distinction amid the religious and the corporeal or the material, the globe and that soul or the strength or the sentience subsist not only in the humans, however, also in the animals, the flora, the rocks, the geographic appearance, for example, the mountains or the rivers or even other bodies of the ordinary environment which include the water nymphs, the vegetation divinity, ranking sprites and many others. Animism can further aspect the life strength to conceptual concepts, for instance, the words, the actual names or the descriptions in the legends. Several of the elements of the non-ancestral globe also regard themselves as animists like the biographer Daniel Quinn, the sculptor Oyekan Lawson and various existing Pagans (King-Tak, 2009).

7.8. FUNDAMENTAL CONCERNS

7.8.1. Environment

The environment is a position where diverse things are like the marshy or even the hot surroundings. It may be living also known as biotic or the non-living also known as abiotic things. It consists of the corporal, the substance, and any other usual forces. The living things exist in their surroundings. They continuously interrelate with it and become accustomed with themselves to circumstances in their surroundings. In the settings, there are dissimilar connections amid nature, the plants, the soil, the aquatic, and any other existing and non-existing things.

Ever since everything is an element of the surroundings of something moreover, the statement environment is implemented to converse about various things. The natives in dissimilar fields of acquaintance implement the word setting differently. The electromagnetic surroundings are the radio impression and any other electromagnetic emission and the magnetic areas (Agarwal, 2005).

In the psychology and the medication, an individual environment is the natives, the physical substances and positions that the individual lives with. The surroundings shape the enlargement and the growth of the individual. It shapes the person's manners, body, intelligence, and compassion.

7.8.1.1. The Connection Among the Humans and the Surroundings

The connections among the humans and the surroundings tend to be mutual. Human beings are twisted by their surroundings and the biological contexts, except they have a big blow to these settings and landscape. This is the sole aim of the underlying prototype that stretches transversely all of human accounts of slavery. Majorly, one may say that the existence is not even achievable without the accurate environmental requirement, but even further than that, the nation tends to be formed by their ecological necessities. If we can observe towards the antique World, believe the responsibility of the masses also the marine in determining the city development of the primeval Greece. As the subsequent example, deem Ancient Egypt, that historically lingering by the side of the Nile canal delta. Believing the issue of the geographic obstacle and the natural limitations, also how their existence or the non-existence have created an effect the frequencies of the assault and the relocation between dissimilar geographical areas. From the primeval

World all the manner throughout to the current, the surroundings have been dangerous in influencing human ethnicity. On the further hand, as narrated before, the bond among the people and the surroundings operates in both courses. The elongated-term implemented in agriculture, in majority of the circumstances, has resulted in the ongoing exhaustion of the soil superiority. We may glance towards instances such as the damming of watercourses or the defrayal of the forests and equally to see the individuals transforming the natural features in which they are inherent in. We may glance towards the collision of industrialization and towards the modern troubles of the ecological dilapidation. In diminutive, human creatures have depended on the surroundings and have been fashioned by the settings; nevertheless, they've twisted those surroundings also (Agarwal, 2005).

Some individuals refer to themselves as environmentalists. They consider we should protect the normal environment, to maintain it safely. Stuffs in the natural surroundings that we consider are known as the natural assets. For instance; the fish, the insects, and the forests. These are the renewable assets because they arrived back normally when we implement them. The non-renewable assets are vital things in the setting around that are restricted for paradigm, the ores, and the fossil energy. Some stuffs in the normal surroundings may kill the creatures, a good example includes the lightning

7.9. THE ENVIRONMENTAL PREDICAMENT

7.9.1. Environmental Moral Codes and Its Inhabitants

Environmental moral codes are the theoretical discipline that deems the ethical and principled connection of the human creatures to the surroundings. In extra terms: what, if several, moral compulsions does man encompass to the maintenance and the trouble of the non-individual globe? While the ethical problems concerning the surroundings have been discussed for centuries, the surrounding ethics did not appear as a theoretical regulation until the year 1970s. Its materialization was the consequence of the increased responsiveness of how the quickly growing globe inhabitants were impacting the settings, also the surroundings consequences that appear with the growing implementation of the pesticides, the knowledge, and the manufacturing (Jardins, 2012).

Environmental moral values help describe man's ethical and principled responsibility toward the surroundings. Nevertheless, human principles become an aspect when glancing at environmental beliefs. Human standards are the stuff that is important to folks that they subsequently implement to evaluate the events or the proceedings. In extra words, human beings assign worth to particular things and then implement this to allocate the values to create decisions concerning whether rather is accurate or mistaken. Human standards are exceptional to each person because not everybody places the identical importance on all the fundamentals of existence. For instance, an individual living in scarcity in an emergent country can discover it morally suitable to slash down the wooded area to create space for an arable farm where he may grow provisions for his folks. Nonetheless, an individual in a growing country may discover these deeds as morally intolerable because the obliteration of the wooded area amplifies carbon dioxide release into the ambiance, which may unenthusiastically impact the surroundings. The environmental moral standards, beside human standards, make for the demanding philosophical deliberate about the human beings interface with the surroundings. Water and the atmosphere pollution, the exhaustion of the usual assets, defeat of biodiversity, the demolition of the ecological unit, and the global atmosphere change are all elements of the surroundings ethics argument (Jardins, 2012).

7.9.2. Environmental Moral Values and Toxic Waste

Environmental effluence played a core responsibility in the initiation of the environmental association in the period of the twentieth century. Even as some ecological ethicists have been concerned that fear about the pollution replicate a moderately shallow shape of environmentalism paying attention on the apprehension of the affluent, toxic waste which is also a momentous threat to the majority of the disadvantaged clusters, the natives of the low-income states, and the non-human being organisms. The ethical concern occurs both in the path of scientific investigation to identify destructive pollutants and in the strategy verdicts about how to normalize them. The latest strategies are required for the generating methodical research that sufficiently reflects the apprehensions and the priorities of the community as an entire relatively than the standards of the polluting productions. The imaginative guiding principle methods are also desirable so as to facilitate we may identify the destructive contaminants which are more speedily and evade them in the production in the first place (Jardins, 2012).

7.9.3. Are Our Ecological Crises a Catastrophe of Ethics?

A biological crisis takes place when revolutionizing the settings of a genus or the inhabitants destabilizes its constant survival. A tiny figure of possible basis consists of the: deprivation of the abiotic biological factor for instance, the enlargement of high temperatures, are less momentous rainfalls amplified stresses from the predation. The Environmental catastrophe is in fact a crisis of realization. Majority of the natives know the ordinary planet is facing enormous challenges and deprivation, however, few recognize the actual extents of the transformation and the deprivation the surroundings facades and its extensive effects on individual interests and for all the other existence on Earth. The effects on the environmental crisis have numerous impacts on the Human physical condition. The Human physical condition can be at the receiving end as a consequence of environmental deprivation. Areas that are uncovered to the toxic atmosphere pollutants may cause respiratory troubles such as pneumonia and asthma. A very big number of individuals are acknowledged to have conked out owed to the meandering effects of the atmosphere contamination. Water contamination, resources catastrophe, the gender unevenness, the effluence, the land toxic waste, the urban extensive, the deforestation, the over invention are several of the common illustrations of the dangerous things that are the basis of overpopulation. The environmental crises can be dealt with by implementing reprocessing the behavior into our everyday existence which is the sole of the majority of the effective methods to help diminish the landfill dissipate, the conserve usual resources, to save territories, to eliminate the pollution, to slice down on the vigor expenditure, and the sluggish down of the global humid.

8

World Trade Organization and Environmental Issues

CONTENTS

8.1. INTRODUCTION

The growth and expansion and development of world trade have brought about the question whether there exists a relationship between trade and the environment. What effect does trade have on the environment? Goods produced by different countries can either be exported or imported and the production of these goods can have an effect on the environment. Another question is whether these effects will increase in the event that trade expands and how they will affect the countries that either import or export the said goods or if it will have a global effect or impact. *Whose responsibility is it to attend to these environmental promises that result from trade agreements?* A question often asked with very little attention received from invested parties in the recent years.

There is a significant relationship between trade and the environment, especially agricultural trade. Sustainable agriculture and rural development and environmental issues are a growing concern and a subject of debate. There is a concern about the relationship between policy measures that are taken to protect the environment and agriculture concerns and the World Trade Organization's policy and regulation on agricultural trade liberalization. As countries become increasingly involved in global trade, international policy frameworks that govern how different governments develop policy are evolving (Tamiotti and World Trade Organization, 2009).

The issue of trade and the environment first received international attention in the year 1991 when there was a challenge between the American and Mexican government when the US banned trade of Tuna fish from Mexico. The argument of the US on this specifically came from and was based on the US marine protect act which prohibited methods of tuna fishing methods that led to the death of dolphins in large numbers therefore banned the importation of tuna from countries that employed such methods. The Mexican government argued that the U.S had violated the General Agreement on Tariffs and Trade (GATT) principle on free trade created by the World Trade Organization and agreed upon by member countries, countries are only allowed to restrict importation in the event a country is trying to protect the safety and health of its citizens and animal protection was not included. Although Mexico did not go forward with fighting the decision of this ban, the situation brought forward the issue of trade and the environment since the tuna issues had been a topic of debate the World Trade Organization foresaw other similar issues in the future related to the environment such as the general protection of the environment, ozone depletion, toxic wastes especially from industrial production, forest protection and climate change all which have a link to international trade.

Agreements made from conventions such as the earth summit of 1992, the WTO agreements of 1994 and the world food summit of 1996 are some of the multilateral environmental agreements at international level that influence the policymaking framework of member states. These agreements will influence policies in environmental protection, trade, and sustainable development. Despite the agreement other parties outside are in disagreement with the measures with regards to trade, agriculture, environmental protection. It is still necessary for the global community to address the challenges they face. A debate related to trade and sustainable development brings forward the issue of poverty, food security and social wellbeing, natural resources management and pollution control (Tamiotti and World Trade Organization, 2009).

To understand this issue, we need to examine the intricacies of trade at both national and international trade. In theory, trade is a way and opportunity for countries to improve their economies by increasing wealth but the expansion of trade to an international level when done practically can have adverse effects on the environment globally. At national level, economic policies are responsive to impacts made to the environment and in simple terms, a country will develop economic policies that take into account how they influence other externalities. In the cause of global trade, there were no set regulations that took into account how expansion of trade beyond national boundaries affected the environment as the authority to create these regulations existed at national level. International trade can create problems when there are transnational environmental effects and impacts. Therefore, the issue of trade and environment should be an issue of concern at both national and international level.

We can create wealth and grow economically with an unchecked use of natural resources. *But at what cost?*

8.2. TRADE AND ENVIRONMENT

8.2.1. What Is the Relationship Between Trade and the Environment? Is Sustainability in Trade and Environment Possible?

The increase in the integration of global value chains and global trade expansion raise questions about their interaction. We try to understand the direct or inverse relationship between these two components. What effects does trade have on the environment? How does the changing environment affect trade? Does liberating trade affect the environment? Are there any long term or short term benefits consequences that result from combining policies that are optimal to the sustainability of both trade and the environment?

Trade expansion results in economic growth which can directly impact the environment by resulting in pollution, natural resources degradation, and biodiversity loss and for some countries; it can lead to intensifying policies related to pollution especially by environmental scientists by application of stringent measures that are environmentally friendly. On the other hand, for well-structured government trade liberalization can be beneficial to the environment, when economic growth, social welfare, and development are supported a country can have the capacity to effectively manage the environment and its resources. It can also create access to technology that increases the efficiency of local industrial and agricultural production by reducing the usage of production inputs such as energy, water, and environmentally harmful substances (Calster and Prévost, 2013).

The consequences of climate change can also affect trade such as rising sea levels and extreme weather conditions and thus the supply, transport, and distributions of products of trade is affected. Most transnational trade is dependent on maritime shipping which accounts to about 80% of trade functions and this can cause delays in shipment. Extreme weather conditions can also decrease productivity by disrupting production activities, for example, they can lead to loss of outputs, destruction of natural resources some of which are used as inputs and increased production time consequently affecting global trade (Calster and Prévost, 2013).

How can environmental and trade policies be optimally combined? At local, regional, national, and international level there is the need to develop environmental policies and institutional frameworks that are effective. A country's welfare is dependent on whether there are policies for environmental protection to cushion the impact of trade liberalization. For a minute let's talk about the trade advantage with trade liberalization having stringent environmental policies is an advantage if the country you're exporting goods follows or has similar stringent environmental policies using the ban of tuna fish importation from Mexico to the US as an example of the Mexican had used appropriate tuna fishing methods a ban would not have been placed on their goods costing them a trade deal which obviously had an economic impact.

The WTO an organization dealing with world trade is working on building a multilateral framework to support transnational trading and at the same time prevent discrimination in the trading process. Inclusion of environmental regulations and policies in trade agreements has been resourceful in harmonizing regulations related to the environment between developing and developed countries. The OECD is also another organization in the forefront in addressing issues related to trade and environment such as environmental and trade agreements that have environmental provisions, stringent policies to foster trade in both environmental goods and services.

The WTO was founded in 1995 but the committee on trade and environment came into existence based on a decision adopted during Uruguay around with the major aim of establishing the relationship between trade and the environment in the promotion of sustainable development. This agenda was extended from the GATT agreement of 1991. Sustainable development and environmental preservation and protection are some of the most principle goals of the World Trade Organization. These goals are found within the Makkaresh agreement which was developed by the WTO and complement the objective to remove and reduce trade barriers and in international trade facilitate the removal of discriminatory trade barriers. WTO does not have a specified environmental agreement but under its rules, members are required to take up trade-related measures that have a protective effect on the environment. Trade openness as one of their objects contributes positively to environmental protection and preservation and also through its mechanism of enforcement and rules, through work in different WTO bodies, and continued efforts under the Doha agenda for development. Within the Doha, agenda there are specific trade negotiations and some assigned tasks to the regular committee for trade and environment (Calster and Prévost, 2013).

8.3. THE WORLD TRADE ORGANIZATION

The World Trade Organization was founded on 1 January the year 1995 with a purpose to reduce trade tariffs and barriers to trade. Its headquarters is centered in William, Geneva Switzerland, and has 164 member states. The

World Trade Organization became functional in 1995 under the Marrakesh agreement signed in 1994 and it substituted the GATT agreement of 1948. The World Trade Organization or, WTO, is an international organization that deals with trade rules between countries worldwide. The main goal of this organization is to ensure that trade activities run freely, smoothly, and in a predictable manner. At its core are trade agreements that are signed off by nations taking part in trade globally and ratified by legislators in the individual countries. The World Trade Organization is headed by a Director-General and currently Roberto Azevedo who is in charge of 600 staff.

The main function of the World Trade Organization includes the administration of world trade agreements, provision of forums for negotiations related to trade, handling trade disagreements, it can function to monitor trading policies for member countries, capacity building as they provide technical help and train developing countries with the aim of assisting them in trade policies and they ensure there is cooperation with other organizations globally. WTO has been involved in establishing a system of agreement that regulates trade. WTO is also a center for economic research and analysis. Regular assessments of global trade are published in its publications and research reports on different topics determined by the organization (Calster and Prévost, 2013).

Three major trade treaties include the GATT dealing with trade goods and the general agreement on trade in services (GATS) and another agreement is on trade-related aspects of intellectual property rights (TRIPS).

The World Trade Organization also deals with governing the internet and policy related to digital issues which is inclusive of e-commerce, arbitration, liberalization of markets related to telecommunication, and trade in communication services. The WTO also organizes public forums yearly bringing together governments, non-governmental organizations, business, and people in academia and businesses to discuss a variety of trade-related topics and issues inclusive of the digital economy (Calster and Prévost, 2013).

Studies done have comprehensively shown that WTO as an organization has boosted trade and has become a great influence to trade agreements (Figure 8.1).

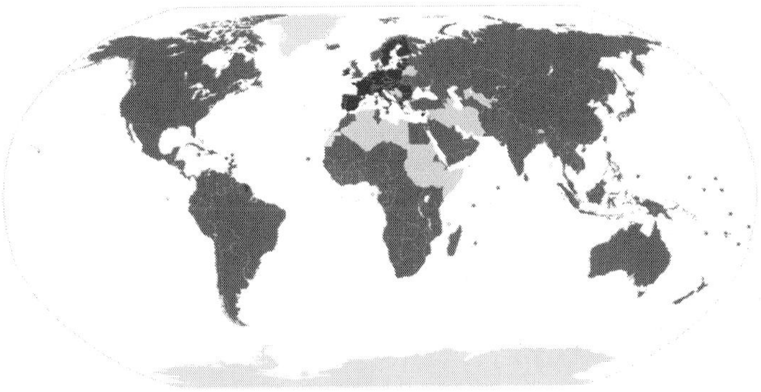

Figure 8.1: Map of World Trade Organization members and stakeholders.

Source: https://en.wikipedia.org/wiki/World_Trade_Organization#/media/File:WTO_enlargement.svg.

8.3.1. History of WTO

The GATT was a treaty established after the Second World War to reduce tariffs and was signed by 23 countries in the US. A comparable organization is the international trade organization. Several negotiations and agreements have taken part under the GATT. The second was the Kenny round in the 1960s brought up the antidumping agreement and some part on the development. The Tokyo round was in the 1970s of which had the aim of tackling trade barriers.

The next was in Uruguay turned the previous agreements into multilateral commitments. It is the agreement that finalized the GATT since it was straining to adapt to the global economy. The GATT is still in existence as a WTO umbrella treaty for trade in goods. The next is the Marrakesh treaty of 1994 which has six parts the agreement to establish WTO, multilateral trade agreements, agreements on trade services, and agreements on intellectual property rights, settlement of disputes and government reviews of trade policies (Tamiotti and World Trade Organization, 2009).

The Doha round which is active up to present. This round of negotiations was done to make trade more inclusive and prevent discrimination of the world poor by slashing trade barriers. Some of the emerging issues during these negotiations were as a result of the differences between developing and developed countries specifically to do with tariff and non-tariff barriers and the results from these negotiations are still uncertain.

8.3.2. The Ministerial Conference

WTO has a decision-making body and at the highest level, there is the ministerial conference. They meet every two years and bring together all WTO members which are countries or custom unions. They take decisions on all trade agreements. Some of the meetings include the inaugural ministerial conference, Cancun conference of 2003, Seattle conference of 1999 and the fourth was the Doha conference of 2001 while the 12th is to be held in Kazakhstan in June of 2020.

8.3.3. The Structure of WTO

The General Council has some of the following different bodies that oversee committees in different areas:

- Trade in goods council. 11 committees exist under their jurisdiction and which specific tasks are assigned. All members participate in the committees although the textiles monitoring body but still fall within the committees. It has a chairman and 10 members. It also has several groups related to textiles.
- TRIPS council. Their functions are related to intellectual property right information, the TRIPs council official records, news, and activities and the details to do with other international organizations in the field for trade services.
- The council for trade services. The general council guides the operations of the council of trade services. The council of trade services oversees the function of the GATS. The council is open to all members.
- The trade negotiations committee. It deals with trade talks that are current or in negotiation. It is chaired by the director-general. Their most recent task was the Doha round which was done in 2012.

The service council is the second council; it has three bodies under it. They include financial services, GAT rules, domestic regulations and commitments that are specific. The council provides services for different committees, working groups and parties. These committees on trade and environment; trade and development; regional trade agreements; restrictions on balance of payments; budget; administration and finally finance and administration. There exist working parties on accession. Trade, debt, and finance: trade and technology transfer (Figure 8.2).

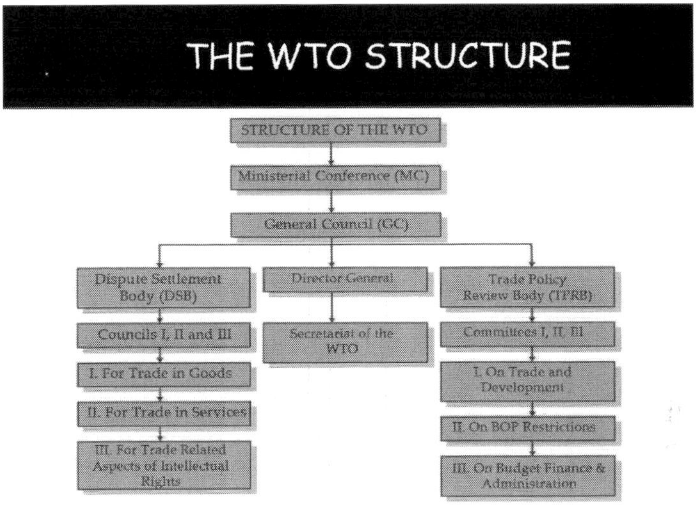

Figure 8.2: Organogram of the World Trade Organization.

Source: https://www.slideshare.net/tinaazwadia/the-world-trade-organization-wto.

Objectives of the World Trade Organization:

- To favorable reductions in tariff and non-tariff are made.
- To eliminate and discourage discrimination in international trade relations.
- To contribute to higher standards of living, employment, growth of income and demand. Promote increased production and trade in goods and services for member countries.
- To positively affect trade by ensuring even the most undeveloped country secures a part in growth as a result of international trade.
- To facilitate the use of world resources in a sustainable and optimal manner for sustainable development.
- To promote an integrated trading system that is more durable and viable with incorporation of the trade resolutions from the Uruguay round of trade negotiations.

Decisions within WTO are made by member governments and the negotiations made for a basis for the rules. WTO sees through votes in the event members cannot reach a consensus. The process of decision making is dominated by the process of decision-making.

8.3.4. Dispute Settlement

The dispute-settlement system of WTO resulted from the evolution of rules that begin with GATT agreement. The final act signed in the Marrakesh agreement of 1994 was based on the agreed-upon rules and understanding in relation to settling disputes and stated the procedures also. It is the core of the multilateral trading system as it is important in the stability of trade agreements between countries. The multilateral system of dispute settling is used in the event a member of WTO is believed to be violating trade rules and agreements instead of solving them unilaterally. The process of dispute settlement involved case-specific panels appointed by the appellate body which is inclusive of the director-general, the secretariat, arbitrators, and advisory experts. Dispute settlement can be through an agreed solution that is mutual and this is to be conducted in a timely manner so that the process does not take more than a year for the panel to make a rule and no more than 16 months if the case is appealed. If the complainant makes the case seem urgent then provisions can be made to ensure the case takes less time. The members have the obligation to accept the dispute settling process as exclusive and compulsory. Despite the process of dispute settlement, the WTO has been unable to settle disputes that result from political disagreements (Tamiotti and World Trade Organization, 2009).

8.3.5. WTO/GATT Principles

The WTO has five basic principles with importance to the pre-1994 GATT and the WTO. As much as WTO develops frameworks that guide trade, it has no responsibility over the trade outcomes between nations. The major role of the World Trade Organization is setting the rules and trade regulations. The five principles include:

- **Nondiscrimination:** It is composed of two components: the first is the most favored nation rule and the national treatment policy. They are found within the WTO rules on goods, services, and intellectual property rights. The scope and nature of these rules differ across the above areas. The MFN rule dictates that a trading WTO member applies the same rules to all trading partners which at the same time should be favorable and therefore there is no room for granting people special favors. The national treatment policy states that imported goods should be treated the same as local goods therefore not discriminating against imported goods.

- **Reciprocity:** It reflects the need to limit or deal with issues that may arise from the MFN rule and free-riding and a need to get better access to foreign markets. In negotiations between trading countries. It is necessary that the gain from these negotiations be greater than that of unilateral liberalization. Reciprocal concessions ensure materialization of these gains.
- **Transparency:** It is required of members to publish their trade regulations to allow for the review decisions that affect trades and provide members with information about members on request. The members are also required to notify WTO in the event of policy changes in relation to trade. They also function to improve predictability and stability by discouraging trade measures such as the use of quotas which set limits on the quantities of imports
- **Binding and Enforceable Commitments:** Enumeration of trade commitments made by WTO members in multilateral is done in a schedule of concessions. These concessions are ceiling bindings and a country can decide to change their decision but after the country has negotiated within their trading partners, which also mean a trading partner may be required to offer compensation for losses. If the process is not satisfactory, the complaining country may invoke the procedures of dispute settlement by the World Trade Organization.
- **Safety Rules-Government are able to Restrict Trade in Some Circumstances:** WTO agreements allow members to take up measures that protect the environment, animal, plant, and public health. In this direction, there are three provisions made. They include attaining economic objectives are attained by articles allowing the use of trade measure, two articles that facilitate fair competition members are not allowed to use environmental protection measures as a means of covering protectionist policies and the third are provisions on the allowing interventions on trade for economic reasons. This is an exception to the MFN principle that permits preferential treatment in relation to developing countries, regional free trade areas, and custom unions.

8.4. GENERAL AGREEMENT OF THESE TRADE SERVICES GATS

8.4.1. Intellectual Property Rights

The agreement on TRIPS. It is an international legal agreement between member nations. It is responsible to set up minimum standards that regulate forms of intellectual property for different governments. Intellectual property rights agreements were included in the Uruguay round after intense lobbying by the United States, The European Union, Japan, and some of the developed nations. The United States' strategy to link intellectual property rights and trade can be traced back to the senior management campaign in the early 1980s in Pfizer. *Who was involved in the mobilization of corporations and maximization of intellectual property privileges and this was the number one priority of trade policies in the United States?* After Uruguay GATT became the basis for the World Trade Organization establishment. To be a member of the World Trade Organization it is compulsory to ratify the TRIPS agreement. Member countries must enact strict intellectual property laws mandated by TRPS if they are looking to obtain hard access to numerous international markets opened by the World Trade Organization and it is therefore the most important multilateral trade agreement and instrument for the globalization of intellectual property rights. States like China and Russia have found the prospect of WTO membership as a powerful incitement since there was little likelihood of them joining the Berne convention (Tamiotti and World Trade Organization, 2009).

TRIPS was negotiated at the end of the Uruguay Round of the General agreement for trade and tariffs between 1989 and 1990 administered by the WTO. For the first time the TRIPS agreement, introduced intellectual property law into the multilateral trading system and still remains the most comprehensive multilateral agreement. In 2001, developed countries were insisting on an overly narrow reading of TRIPS and this became a concern of developing countries and these went further to initiate talks that resulted in the Doha declaration. This is a statement clarifying the scope of the TRIPS agreement and states that the TRIPS agreement should be interpreted in light of its aim in the promotion of medicine access to all people globally.

The WTO, specifies the need for members to provide copyrights, covering authors and copyright holders as well as holders for related rights such as performers, sound recording producers, and broadcasting organizations; Geographical indicators, industrial designs, integrated circuit layout designs, patents, new plant varieties, trademarks, trade names and information that is confidential (Tamiotti and World Trade Organization, 2009).

TRIPS specifies procedures of enforcement, solutions, and procedures of dispute resolution. WTO tries to ensure that the TRIPS agreements meet its objectives in contributing to the promotion of advancements in technology such as innovative technologies, transfer, and dissemination of technology and ensuring that producers and users both benefit from these new technologies in a conducive manner to both social and economic welfare and finally to create a balance between rights and obligations.

- **TRIPS Requirements of Member Countries:** It requires its members to provide strong protection to intellectual property rights. For example, under TRIPS
- Terms for copyrights must have a 50-year extension unless it is based on an author's life.
- Copyright must not be based on formality such as registrations but automatically granted as per the Berne Convention.
- Under the copyright law, computer programs must be regarded as literary works and should receive similar terms of protection.
- The Berne three-step test constrains national exceptions to copyright, for example, fair use.
- In all fields of technology, patents must be granted provided they meet the terms for patentability although exceptions can be made for certain public interests.
- Provided that a normal exploitation of the work and normal exploitation of patent are not in conflict then exceptions to exclusive rights are expected to be limited.
- There is no allowance of unreasonable prejudice towards the legitimate interests of the right holders of computer programs and patents.
- Patent rights should take into account the legitimate interests of third parties.
- It has a MFN clause. Intellectual property rights may not offer any benefit to countries that are not signatories of the TRIPS agreement.

8.4.2. Non-Tariff Barriers

Non-tariff barriers or measures to trade are barriers to trade that lead to the restriction in the trade of imports and exports through mechanisms other than imposing the normal tariffs to trade. According to the World Trade Organization non-tariff barriers to trade, include import licensing, rules of valuation of goods at customs, inspections during shipment, rules of origin, and prepared investment measures on trade.

Nontariffs are similar to tariffs with an exception of export subsidies and quotas. During the eight rounds of negotiations in the GATT by the World Trade Organization tariffs for producing goods were reduced. After tariffs were reduced, the principle of protectionism brought up the need to introduce newer non-tariff barriers such as trade barriers that are technical. The increasing popularity of technical trade barriers have been impacted on by the increase in demand for environmentally friendly products. Many non-tariff barriers are controlled by agreements from the World Trade Organization. Which were as a result of the Uruguay negotiations that are the technical trade barriers, SPS measures, textiles, and clothing agreement and also the GATT articles. Non-tariff barriers are protectionist measures unless they are related to externalities and information asymmetries which can be difficulties in the market. This is between producers and consumers. Non-tariff barriers are the new form of protections which is a replacement for tariffs. Some of the best examples that would easily explain these are safety standards and labeling requirements. Most industrialized countries find the need to use nontrade barriers in relation to industries that need sensitive imports, as well as other restrictions (Tamiotti and World Trade Organization, 2009).

There are different types of non-tariff barriers such as:

- **Policies that are for Protection:** These favor firms and enterprises at the expense of other nations. Examples of these tariffs include quotas, anti-dumping laws, practices for public procurement, and requirements for local content. The consequences of applying these policies include challenges brought up at the World Trade Organization conventions and other trade organizations and forums.
- **Policies for Assistance:** These policies favor firms and enterprises but unlike the protectionist policies, it is not done at the expense of other countries. They include bailouts from industries, and subsidies that are domestic in nature.

- **Policies that are Non-Protectionist:** They are for the protection of peoples and animal's health. They also protect the environment. They include licensing, packaging, and labeling requirements, rules for sanitation, inspection of plants and animals, bans can be imposed on imported goods that are of questionable harvesting, fishing, or production.

Other variants to the types of non-tariff barriers classification include its division in internal taxes, barriers in administration, regulations on health and sanitation, and policies on procurement. Others classify them based on limitations in trade, custom, and entry procedures, product standards, and participation of the government in trade. Examples of non-tariff barriers include border delays like that of Turkey and the European Union despite them having a customs union the export of Turkish goods are subjected to overheads that expose the Turkish traders to losses that cost them an equivalent of 3 billion Euros every year.

- **Embargoes:** Prohibition of trade imposed on certain goods such as quotas in relation to imports from certain countries especially for sanitary conditions, sometimes the reasons are political. They are generally considered illegal.
- **Controls on Foreign Exchange:** It can include manipulation of exchange rates by states or limitation of the foreign currency supply.
- **Import Deposits:** Is a fee paid by an importer to a central bank for a given period of time and may be equal to cost of importation or part of it.
- **Licenses:** A lot of countries apply this non-tariff method. This requires that a state gives permits in the form of licenses to carry out transactions while importing or exporting trade goods. A license as a trading instrument in foreign trade is based on standard agreements and it also includes provisions made by the GATT by the World Trade Organization an example of agreement are procedures for import licensing.
- **Standards:** These are mostly used in the protection of safety and health of locals and also to protect the environment. Some of the standards imposed are on labeling, classification, and product testing to ensure they meet domestic standards to restrict product sales to ensure they meet or exceed the set standards.

- **Quotas:** This applies to certain goods for a time period and also is referred to as quantitative control. They limit import or export of goods especially foreign trade. These barriers have a cost implication and the quantity of products that can be traded the logic behind it include price manipulation, guaranteeing supply in times of shortage and the strategic control of goods.

8.4.3. Subsidies and Countervailing Measures

This agreement addresses two closely related but separate subjects. The first is the multilateral discipline that regulates the provision of the subsidies while the second is the use of countervailing measures that offset injury caused by subsidized imports. They govern whether a member can provide subsidy on trade goods or services. The WTO dispute settlement mechanism enforces these rules through invocation. Countervailing duties are an instrument that is unilateral which members can apply after the member investigates and determines that the set forth criteria in the SCM agreement are satisfied.

8.4.3.1. The Agreement Structure

The first part of the agreement only applies to subsidies provided to an industry or enterprise. The second and the third part break down the agreement into two categories, prohibited, and actionable, and also establish rules and procedures related to these categories. Part v of this agreement provides requirements that are procedural and substantive to be fulfilled before application of countervailing measures on subsidized imports by a member state. The next two parts establish surveillance modalities for SCM agreement implementation. The last part contains rules on special and differential treatment for categories of members from developing countries. The last two parts have final and dispute settlement provisions (Tamiotti and World Trade Organization, 2009).

8.4.3.2. Coverage of the Agreement

The first part of the agreement explains the meaning of subsidy and specificity. Subsidy in the agreement has three basic elements. They include:

- A contribution financially;
- By a government and public body; and
- Which holds a benefit.

The elements must be fulfilled for a subsidy to exist. For a financial contribution to exist it must be made by a government or a public body by a member within the territory and also be beneficial.

Specificity-assuming a trade measure is a subsidy based on the SCM agreement it is not subject to the agreement unless it has been specifically being provided to an enterprise or an industry. If it distorts resource allocation, it should be subject to discipline. Only specific subsidies are subject to the SCM agreement disciplines. There are four types of specificity within the agreement. They include the enterprise, industry, regional, and prohibited subsidies.

Subsidies are categorized into two prohibited and actionable all specific subsidies fall into these categories. Two of the subsidies that fall within this category are prohibited by the World Trade Organization export subsidies and local content subsidies. Actionable subsidies are not prohibited but are subject to challenge. They can be settled through countervailing action if they cause adverse effects if it causes injury to the domestic industry, if serious prejudice arises from these effects, impaired benefits from the GATT agreement of 1994 assuming the improved market access is affected by subsidization (Calster and Prévost, 2013).

8.4.3.3. Countervailing Measures

Part V of the SCM agreement brings forth requirements that must be met in order to impose a countervailing measure. The requirements can either be substantive or procedural. If a member fails to respect these requirements it can be considered as a dispute settlement and this may for a basis for invalidation of the countervailing measure. Substantive rules state that countervailing measures cannot be imposed unless it is determined that there are subsidized imports, a domestic industry injury, and a causal link between the injury and the subsidized imports. Criteria regarding injury and causation are found in part v. it also contains rules with regards to the existence and amount of benefit. The objectives of these rules are to ensure that investigations are transparently conducted. They ensure that all involved parties have an opportunity to defend their interests. There are important changes in the agreement including standing, a preliminary investigation, sunset, and judicial review (Calster and Prévost, 2013).

8.4.4. Imposing Trade Restrictions

Trade restrictions can be placed by a government especially on international trade. The theory of comparative advantage explains how trade barriers can decrease an economy's efficiency. Most trade barriers have a similarity in the way they are imposed, most of them have a cost implication such as money, time, bureaucracy, and quotas. A trade conflict can result from countries using trade barriers against each other. Tariffs impose an extra financial burden while a non-tariff uses overt or covert means to restrict trade. Free trade, in theory, means the removal of these trade barriers to make trade more efficient in the exception of those that protect health and safety.

The impacts of imposing trade restriction mostly affect small trading firms for one they affect free trade and also result to limited product choices in the market while removal of trade barriers can also negatively affect trade by market distortion especially when developing countries overproduce goods and dump them to developing countries at low prices causing disadvantages to the farmers from developing countries. Article XI of the GATT provides provisions for regulating quantitative restrictions. It includes prohibitions other than taxes or tariffs applied or maintained by the World Trade Organization member countries which are made effective through quotas or licenses. These restrictions are allowed in certain circumstances. Such prohibitions are related to the trade of drugs, weapons, and measures to protect the environment. In some cases these restriction result from obligations required of countries outside the WTO framework, for example, in relation to the Montreal protocol on substances that deplete the ozone layer or CITES convention to protect endangered species (Calster and Prévost, 2013).

The extent of the measures is found within this article and notified to the WTO and can be justified under this. The members are obligated to ensure that these measures are not discriminatory included in the XIII article in the GATT. This is to be notified at regular periods. A notification is required to be submitted if such a measure is imposed and is adopted by the council of trade. The large majority of these measures have provisions provided in the 1994 GATT agreement:

- Article XX b-measures to protect human, plants, animals, plants, and health;
- Article XX g-conservation of the environment;
- Article XXI b protection of security services that is essential.

8.4.5. Trade Without Discrimination

Trade without discrimination is under the most favored nation clause in article 1 of the general agreement on trade. The World Trade Organization member states are committing to giving similar treatment to other members regardless of where they come from. The most favorable nation principle means members are not allowed to discriminate against other members but some special conditions are an exception. If a member country provides special treatment to another country, it is required to grant special treatment similarly to other countries this allows equality. It therefore one of the most important principles developed by the World Trade Organization. It is important in the GATT that controls trade of goods and it is also a priority in the GATS and also the agreement on intellectual property rights. Each agreement covers the principle differently.

National treatment; equal treatment of foreigners and locals-trading goods in both cases should be treated equally at least after they have both reached the market. This should also apply to local and foreign goods, trademarks, copyrights, and patents. This principle is found in all the three main trade agreements of the World Trade Organization. In article 3 of the GATT, article 17 of the GATS and article 3 of the TRIPS agreement. It only applies once the goods or services have reached the market. Therefore, some aspects such as custom duty do not violate the principle or not charging similar taxes on locally produced goods.

8.5. TRADE AND ENVIRONMENTAL MILESTONES

The relationship between trade and environment is an issue that is not new. Their link has an impact on policies related to the environment. This was recognized as early as the seventies. The concern of these impacts on social and economic growth led to calling of the Stockholm conference on how to manage the human environment. This conference was held by the United Nations organization and the secretariat for the GATT agreement was asked to make a contribution and they prepared a study on industrial pollution and international trade leading to creation of the green protectionism (Calster and Prévost, 2013).

In 1971 the EMIT group was created in relation to GATT by council members, the environmental measures and international trade and they were to convene upon request by member states and it was not until 1991 that the group was asked to convene upon the request of the European free trade association. The reason for this request was related to the 1992 United Nations conference on environment and development where the group was asked to contribute and there were also new developments in the 20 years after its formation (Tamiotti and World Trade Organization, 2009).

8.5.1. Developments Between 1971 and 1991

Between these periods, environmental policies began having an impact on trade with increased trade flows and the effect had become more widespread.

- Tokyo round of trade negotiations 1973–1979 they included talks on how environmental measures would affect trade. The technical barriers to trade were negotiations.
- The Uruguay round of negotiations 1986–1994 modifications were made to the technical barriers to trade agreements and some modifications were made to environmental issues which were addressed in the GATT agreement, SPS agreements and the TRIPS agreement
- In 1982 the GATT agreement ministerial meeting on measures to bring to control products prohibited domestically on the grounds of bringing harm to human, animal, plant life or health and also to the environment. This led to creation of a group to work on export of domestically prohibited goods and other hazardous substances.
- In 1991 the conflict between Mexico and the United States was brought up as Mexico disputed the embargo against them by the United States and the panel ruled in favor of Mexico. The ruling was heavily criticized by environmentalists and other environmental groups who argued that the trade rules did not consider environmental protection.
- In 1987 the world commission on environment while discussing issues on development in the report, one of the key problems identified as causing degradation to the environment was poverty it was argued that economic growth and development facilitated by international trade would also lead to generation of necessary resources to combat environmental degradation.
- In 1992 at the Rio conventions more developments were made after the activation of the EMIT group saw the adoption of Agenda 21 which addressed the importance of promoting sustainable development through other means and international trade. This discussion led to linkage of trade and environment that was to be

signed in 1994.

- The Marrakesh agreement signed in 1994 worked towards achieving sustainable development with the view of improving living standards while concurrently allowing member countries to make use of resources optimally in accordance to the objectives of sustainable development that work towards combating environmental degradation and preserving it. The agreement also makes the recognition that sustainable development is an important part of the trading system and shows that the World Trade Organization places an importance to protecting the environment.

In 1994, the ministers signed a decision on trade and the environment and the document states that there shouldn't be or no need be any contradiction in policy between upholding and safeguarding a multilateral trading system that is open, non-discriminatory, and equitable. At the same time acting to protect the environment and still promoting sustainable development on the other hand. This led to the developing committee on trade and environment that is open to indiscriminate membership. It also includes observers and international organizations. It has a broad mandate; they are tasked with understanding the trade and environment relationship in order to contribute sustainable development.

The importance and contribution to trade in relation to the environment have also been recognized in the Johannesburg summit 2002 and the UN world summit 2005.

There are different environmental disputes handled by the World Trade Organization. Under the GATT agreement, several disputes were handled. The United States banned the import of tuna products from Canada and a ruling was adopted in 1982. In 1988, a ruling was also adopted for the case against the United States by Canada on exports of unprocessed herring and salmon. In 1990, another case was brought by the United States against Thailand and others up to 1994.

Under the World Trade Organization, several environmental and health-related disputes have been handled. In 2001, a case brought by Canada against France on banning products containing asbestos and Canada's challenge was rejected. The challenge in the way WTO handles disputes are the fact that the trade organization sometimes favor's trade over environmental concern.

8.6. RECONCILING TRADE WITH ENVIRONMENT

Trade brings harm to the environment in different ways. Measures to improve trade such as liberalization may lead to depletion of natural resources that are used as inputs and also causes environmental pollution. Opening borders to international corporations might cause companies to migrate to pollution havens. Hence undermining high environmental stands of countries hosting them. The third issue is that environmental issues can be undermined in dispute settlement prioritizing trade instead and in simple terms, it can have a biased nature. Constructive suggestions can be made and can be reconciled in the future when trade negotiations are in place. They include the use of environmental organizations to use their influence to bring adequate policies that can prevent or mitigate environmental degradation and use to influence decisions made by the World Trade Organization members to influence environmental management.

Regular environmental assessments (EAs) of the effects of agreements related to trade should be done on a regular basis and agreed upon. The effects of this strategy will have a positive impact if they press on liberalization of trade that actually positively affects the environment and accompanying these with adequate policies related to agriculture, fisheries, and the energy sector (Tamiotti and World Trade Organization, 2009).

To strengthen the outcome of disputes related to the environment, in addition to panelists from member countries there is need to also include members from environmental organizations and also solve these disputes in the presence of the public rather than being solved behind closed doors. On the other hand, environmentalists should not press on environmental policies that are protectionist in nature and prove to be disadvantageous to some member countries especially those from developing countries or those that compensate for international differences. The call for environmental tariffs can be mistakenly taken as support for protectionist factions who have an idealist or discriminatory basis. They should also try and gain support from developing countries whose priorities are poverty alleviation and creation of wealth over environmental issues (Tamiotti and World Trade Organization, 2009).

8.6.1. The Winnipeg Principles

In an effort to reconcile trade and policies related to environmental protection, the international institute for sustainable development set up principles which were known as the Winnipeg principles on trade and sustainable development. Application of these principles could be of important benefit. It would allow countries to enjoy both prosperity together with an environment that is healthy, since other issues in agriculture, food safety; intellectual property rights are significant to the environment. The principles include

- Efficiency and cost internalization-trade agreements and policies should support efficiency in the use of natural resources and at the same ensure the true value of the resources are reflected in the costs.
- Environmental integrity-creation of multilateral trade systems that contribute to the preservation of environmental integrity considering the capacity of the ecosystem to regenerate, prevention of damage and the protection of threatened areas.
- International cooperation-development of trade systems which positively impact linkages across borders and identities. Improve international cooperation systems.
- Subsidiary-ensuring the decision-making process is done effectively, for example, with regards to trade vs. environmental conflicts.
- Science and precaution-respect of precautionary principles in trade agreements, it can apply to short term situations that lack a scientific base or evidence. Long term decisions can be based on scientific evidence.
- Openness while conducting trade negotiations is necessary and therefore they should be conducted in a manner that allows the participation of different parties and also with transparency.

8.7. ENVIRONMENTAL CONCERNS AND THE WORLD TRADE ORGANIZATION

The World Trade Organization has doesn't not have an environmental specific trade agreement. Their agreements are in line with the government's rights to protect the environment provided certain conditions are met pertaining to the environment. Objectives of sustainable development and protection of the environment are important and therefore stated within the agreement prior to the establishment of the World Trade Organization (Calster and Prévost, 2013).

Emphasis on the importance of environmental policies is a recent concern in the 60 years since the multilateral trading system was established and in the Uruguay round of 95, trade ministers began working on the program on trade and environment by creating a committee. The Doha trade negotiations discussed some of the issues.

The committee is involved in the study of the relationship between trade and the environment and from these studies makes appropriate recommendations about any changes that may be needed in trade agreements. Their work is based on two relevant principles.

- It's only task is to study issues that arise from environmental policies and how these policies affect international trade. This is because it is not an organization whose core function is protection of the environment. Therefore, the agency does not task itself with the responsibility of intervening in national or international environmental policies or creating regulation, policies, or rules that deal with the environment. It leaves these to other agencies or organizations that deal with the environment.
- In the event the trade and environment committee encounters problems, it is expected to resolve these issues in a manner that still upholds the principles of the World Trade Organization trading system. To promote sustainable development, contribute to conservation and protection of the environment WTO members are expected or convinced to practice equitable, open, and apply trade practices that are not discriminatory in nature. This practice was recognized from negotiations and talks from the Rio summit in 1992 and the world summit of Johannesburg on sustainable development. The programs the committee works on are in 10 areas. The World Trade Organization brings up issues important to them that they use to drive the agendas.

8.7.1. Environmental Agreements and the World Trade Organization and Their Relation

How do green trade measures and the World Trade Organization trading system relate?

There are 200 international agreements outside the WTO that are related to environmental issues and they are called multilateral environmental agreements. Approximately 20 of these are related to trade; some set bans on certain products and restrict quantities in trade of other products in some circumstances. Some of these agreements include The Montreal Protocol on Substances that Deplete the Ozone Layer (15 September 1987), The Basel Convention on the Control of Transboundary Movements of Hazardous Wastes and their Disposal (22 March 1989) and Convention on International Trade in Endangered Species of Wild Fauna and Flora (3 March 1973). According to the committee, the principles do not conflict with environmental agreements. It also notes that clauses in the agreements on goods, intellectual property rights, allow governments to prioritize their policies on the environment. Environmental problems can be dealt with through agreements related to the environment (Calster and Prévost, 2013).

The committee also notes that measures taken to protect the environment that have a direct or in direct impact on trade area should be taken into account when making agreements to do with the environment. It also states the fact trade restrictions are not only the solutions to environmental issues, other alternatives can include capacity building of countries by assisting countries to access environmentally-friendly technology, provision of financial assistance and provision of training. Up to now no action that affects trade and taken under the environmental trade agreement has been challenged in the WTO-GATT system. The World Trade Organization committee is mainly concerned with what happens when a country takes action against a country against another country that has not signed the environmental agreement.

In relation to trade disputes, in the case one arises under the environmental agreement outside the WTO taken against trade considering the parties have both signed the agreement, then both parties should try to settle the disputes on grounds of the agreement, the WTO will then provide a forum to settle the dispute. This does not mean the environmental issues will not be given consideration when handling the dispute. The agreements by the WTO allow panel members handling the dispute to seek advice from experts on the environmental issues.

8.7.1.1. Eco Labeling

Some of the environmental policy instruments include the labeling of environmentally friendly products. The world trade program requirements on labeling should not discriminate in trading. Product labeling to describe whether a product is environmentally friendly is still a subject of discussion in relation to technical trade barriers.

8.7.1.2. Transparency

Members of the World Trade Organization are expected to provide information on environmental policies they adopt in order to prevent applying them as protectionist measures especially when they have consequences or impacts on trade. They are required to notify the organization if there are any changes to the policies. The committee states that there is no need to change rules for this purpose. A database is placed for members to access information on policies from different countries.

8.7.1.3. Goods Domestically Prohibited

It is a concern to the majority of developing countries with the worry that some hazardous substances are exported to their markets without informing them on the health and environmental implications. The Basel Convention has provisions to control the transnational movement of hazardous wastes that are hazardous to their disposal. The World Trade Organization plays a complimentary role but does not try to copy it.

8.7.1.4. Trade Liberalization and Sustainable Development

The relationship between trade liberalization was analyzed in the Uruguay round of trade negotiations and environmental protection. The members agreed that trade liberalization would have a positive impact to both trade and the environment but the subject is still under discussion given different views from other parties who claim otherwise by holding a different opinion.

8.7.1.5. Intellectual Property Rights

Discussions between trade and the environment have opened grounds for lengthy discussions since previously there was little understanding on the issue and how trading systems would affect the environment or vice versa. In relation to services there is needed to further understand the relationship between protecting the environment and the GATS. The committee argues and states that the TRIPS agreement can assist countries to gain access to technology that is environmentally sound. More effort is still needed to understand the relationship between the environment and the TRIPS agreement and also on diversity of species in the ecology.

8.8. PUBLIC PARTICIPATION

The engagement of civil society is key to the achievement of sustainable development and environmental goals as mentioned in the 2030 agenda for sustainable development. Despite the mention, the subject of inclusive public participation is not accorded the same attention as other sustainable development goals. There are various reasons for involving the public in decisions related to the environment. The government sometimes cannot reach environmental protection goals alone since they also need the support and guidance from their citizens. This practice improves the quality of decisions made. Top-down policymaking without the opinion of the public can lead to the development of policies that do not project priorities of the local population and specific contexts of development. It can also address distrust that arises from the information based on expert views that dominate decision making. An increase in the participation of the public builds more engagement from its citizens. This means that the decisions made are accompanied by a certain form of legitimacy to them and this ensures that policies made are based on valuable local knowledge. In making decisions related to the environment, the participation of the public includes both the mobilization and the active participation of the civil society. Challenges to the expansion of public participation include limited resources such as time and money. Expansion of public participation can also reflect poorly on the decision making process on the needs of marginal groups such as women and children. They also have the potential to deepen inequalities within a society. Another issue can be lack of transparency in how public participation results affect the policymaking process. Another challenge is the translation of ideal public participation into practice as the mechanisms of enhancing public participation can bring forth inequalities that are already in existence (Tamiotti and World Trade Organization, 2009) (Figure 8.3).

Figure 8.3: Public forum of the WTO in 2010.

Source: https://www.flickr.com/photos/world_trade_organization/5008045662/in/photostream/.

Those making policies must therefore provide resources necessary in the facilitation of both inclusive and equitable participation. They should also ensure that the decision-making process is carried out with openness.

The World Trade Organization is sometimes criticized about the transparency of its decisions and the lack of the involvement of other civil societies allowing both private and public participation. Several authors have tried to assess the potential of public participation on decision making by the World Trade Organization. With trade, liberalization comes with a greater involvement of people in trade and greater impact on health, well-being, and development. It begins by trying to create a framework that indicates the parameters that are significant in the implementation of public participation. Proposals to formalize public participation by the World Trade Organization have the view that it would help to correct the problem of illegitimacy in the decision making process of the World Trade Organization. It is argued that removal of non-tariff barriers should be a subject to public participation some of which pertain to the well-being of people or sometimes the sensitive disputes solved by bodies such as the World Trade Organization can also be subject to public participation. Despite the need for public participation, it is important to understand how the adoption of such a measure would affect the outcomes of decisions and how to regulate public participation. Despite the public not being actively or directly, involved the World Trade Organization maintains regular dialogue with non-governmental organizations to create awareness on their own activities. They also keep open their activities by providing information through their website, social media page, briefings. They also have a forum for debate on issues related to trade and allowance for submission of papers related to their activities and mandate (Tamiotti and World Trade Organization, 2009).

9

Environmental Management Education and Awareness

CONTENTS

9.1. INTRODUCTION

Education refers to the process of receiving or offering systematic instruction that is in most cases given in learning institutions. Awareness refers to the concern about and well-informed interest in a given situation or development. In this case, the subject matter is environmental management. Environmental management refers to the management of interaction and impact of human activities on the ecosystem. This field focuses on ensuring that the ecosystem services as well as biodiversity are maintained and protected to allow equitable use by the future generations. It also places an emphasis on maintaining the integrity of the environment by looking into matters such as the economical, ethical, and scientific variables. It also seeks out to identify factors important in the conflicts that may develop between the protection of the environment as well as meeting the needs of the population. Environmental management is usually a role allocated to a specific department (s) as designated in government charters. For instance, in Kenya, the body in charge of environmental management is the National Environment Management Authority (NEMA). It is also used in other countries like African countries.

Education on environmental management involves organized efforts to teach individuals better ways of managing the environment so that humans can manage their behavior while in the ecosystem with the aim of enabling them to live sustainably. It has involved combined efforts of UNESCO. The organization places most of its emphasis on the role of the EE in ensuring that the future of the development goals is maintained. This is important for the protection of the environment to be maintained to ensure the eradication of poverty, reduction in the inequalities and ensure that the development will occur (Nag, 2005).

The need for education and awareness on environmental management was due to the lack of environmental consciousness among the people. This is a very important (key) role in ensuring that individuals are able to make informed decisions concerning the environment. As the environmental acts involve the public being involved in making decisions involving the environment, the citizens are required to have knowledge on the environment. They need to identify factors that will cause a degradation on the health of the environment and those that will aid in improving the health of the environment. It was established that populations living in the rural areas rely mostly on the natural resources from the environment to conduct their daily activities.

For instance, they cook using firewood, among other fuel sources. As some of them lack knowledge on the environment, some have used these resources such that they have been depleted to some extent. They therefore become the de facto managers of renewable natural provide. They require knowledge on environmental management. Therefore, the existing systems managing the resources are improved and made sustainable; it will improve the status of the environment. There has been development of environmental programs that will require the active participation and support from the communities (Nag, 2005).

As stated above most of the environmental activities are conducted by boards and organizations set up by the government or by citizens willing to assist. In the case of rural areas in a bid to give off environmental education, it has required the use of grass-root partnerships with the village-based non-governmental organizations set up among the people with the aim of promoting environmental awareness and education. There is a great need of promoting environmental education and awareness among individuals. It is also important that all stakeholders, as well as the public, are informed on matters of the environment. They need to be aware that the depletion of the natural resources is similar to the destruction of the prosperity of the future generations. This is because forests will disappear, water will be exhausted or polluted, and the fertility of the land will also reduce thereby causing a limitation in the survival of population groups as the resources will be scarce. The environmental education is conducted with the aim of promoting environmental conservation and the proper use of the natural resources from the local, regional, and international perspectives (Nag, 2005) (Figure 9.1).

Figure 9.1: Environmental education and awareness in a rural community.

Source: https://panorama.solutions/es/building-block/community-environmental-education-and-awareness.

9.2. OBJECTIVES OF PROGRAMS CONDUCTING EDUCATION AND AWARENESS ON ENVIRONMENTAL MANAGEMENT

To increase awareness on the environment by gathering, processing, disseminating, and exchange of information on environmental and development issues. The information obtained is obtained locally or internationally. It is also tasked with supporting the development of organizations and NGOs in the easy response to environmental issues. This is through the use of improved environmental education resources and training activities. This may involve the use of technology. The programs have the task of organizing short courses, seminars, training programs, and workshops that offer knowledge on a variety of subjects that are related to environmental management. They also conduct projects that conduct public awareness campaigns that allow individuals in the villages to acquire knowledge on environmental management.

There are various ways in which education and awareness on environmental education is given off to the public. One of the ways include the offering of such education at the preschool, primary, and secondary levels, at the undergraduate level, workshops, seminars, and also giving the public education.

9.3. EDUCATION AND AWARENESS IN THE PRESCHOOL, PRIMARY, AND SECONDARY LEVELS

This involves knowledge on environmental management being given to children while in the preschool, primary, and secondary levels. This is done through teaching subjects that contain information related to environmental management. Such subjects include science and social studies. The subjects involve the students learning the importance of environmental management and ways they can conserve the environment. They are also taught on the impacts of poor environmental management. For instance, in social studies they learn of iconic people such as Professor Wangari Mathai known for her great work in championing for the protection of the environment. In the primary level, children learn about environmental management through the various activities taught in these subjects such as tree planting (Figure 9.2).

Figure 9.2: Professor Wangari Mathai championing for the protection of the environment.

Source: http://blog.worldagroforestry.org/index.php/2016/03/17/at-launch-the-wangari-maathai-foundation-unveils-major-project-for-environment-and-society/.

As students' progress to higher levels such as the secondary level, they get to learn on human activities that if not properly managed can lead to the destruction of the environment. Such activities include mining, farming, and deforestation. If these activities are carried out extensively without care, it would lead to the scarcity of the natural resources so that the future generations will not be able to access these resources. In such studies, the students are taken in field trips where they can see the impacts of the various human activities on the environment. They are also taken to the organizations that are in charge of environmental management where they are able to learn much about the environment. They also get to know how they handle various issues concerning the environment (Nag, 2005).

9.4. EDUCATION AND AWARENESS IN THE UNDERGRADUATE LEVELS

Education in the undergraduate levels involves individuals being given education on educational management. Most of the universities offer units such as development studies which involve the individuals being taught on various things concerning development on the environment. Development in the long run has its own effects in the environment as it involves the incorporation of various natural resources of the environment in a bid to develop in the environment. As these students are going to be released to the world to be future employees and makers of the nation. If they are equipped with knowledge on how to manage the natural resources to a level that will maintain the available resources.

The institutions offer units that deal with environmental management. Some of these units include forestry, biotechnology, mining, and technology. In higher learning, environmental education is a multi-discipline involving subject such as chemistry, earth science, ecology, and atmospheric science. The

9.5. ROLE OF UNESCO IN ENVIRONMENTAL AWARENESS AND EDUCATION

The United Nations Educational, Scientific, and Cultural Organization (UNESCO) is known for its great involvement in environmental protection through environmental awareness and education ever since it was created. It has led to the creation of the IUCN (International Union for the Conservation of Nature) in 1948. It has now been made to be the World Conservation Union. This was the first NGO whose major mandate is to conserve the natural environment. This board was greatly involved in the Human Environment conference in Stockholm that was held in Sweden in 1972. In conjunction with UNEP, it has led to the International Environmental Education Program. Their mission is to set a vision as well as give off practical guidance on how individuals can mobilize education in a bid to increase environmental management awareness (Nag, 2005).

The organization launched its first environmental newsletter in 1976 known as Connect. This newsletter served as a clearing house that facilitates the transfer of information on environmental education. It generally aims at promoting activities in the IEEP. It serves to create a network for institutions as well as individuals that have shown interest in active environmental education management. The cooperation between the two boards has contributed to the co-organization of four major international conferences focused on environmental education (Nag, 2005).

Some of these conferences include the Environmental education in Tbilisi in Georgina, the International Strategy for Action in the field of Environmental Education and training and Environment and society; Education and public awareness for sustainability. All these conferences had a common highlight on the agenda and that was the importance of good environmental management in sustainable development.

The conference was held in Tbilisi that led to the further exploration of education on environmental matters. UNESCO in conjunction with other bodies made this the first intergovernmental conference aimed at improving education and awareness on matters environment. There were goals laid out in the conference which include creating a clear awareness of, interest, and concern about the economic, political, social, as well as economic interdependence in both urban and rural areas. This has led to the creation of other bodies that assist in the education and awareness on environmental education in the given states (Nath, 2009).

9.6. FOCUS POINTS IN ENVIRONMENTAL EDUCATION AND AWARENESS

There is an emphasis placed in given areas involving education and awareness on the environment. The focus is placed on engaging with the citizens of the nation of all demographics to think ethically, critically, creatively when conducting an evaluation on matters environment. These places need the public to have knowledge on matters environment. They also focus on making educated judgments on the environmental issues. The educated judgment is made using information from the research conducted on the environment. They may include results from scientific research and use of technology in making such decisions.

They also focus on developing skills and commitments to the act (environmental act) collectively as well as independently to enhance and sustain the environment. They focus on improving the appreciation of the environment which may result in the positive change in the environmental behavior of human beings. There have been a number of attributes that have led to the need for environmental management education and awareness. They include the need to enhance real-world problem solving, to strengthen the physical activity and the dieting quality, as well as improve the leadership and communication while working in the environmental groups (Nath, 2009).

9.7. CHARACTERISTICS OF ENVIRONMENTAL EDUCATION

It is concerned with the knowledge of the individual, society, and the universe that has the attempts of creating individual understanding of environmental matters. It focuses on making the individuals be creative and encourages them to take up their responsibility of decision-making being instilled in the human beings. The other characteristic is that it is concerned with the attitudes that people hold for themselves, towards other individuals, towards other groups of individuals and towards their environment. The constellations of this matter form ideas that will to a much extent affect the level of living as well as the quality of time.

Environmental education does not only deal with resource-use, ecology, sociology, philosophy, art appreciation or management but also an interdisciplinary that embraces the humanities, social sciences, science, and technology. It is likened to the measure for the purpose of developing cognitive understanding, change in attitude and belief, and facilitating a motivation for the change in behavior and action.

9.8. FACTS FOR SCHOOL AND NON-SCHOOL ENVIRONMENTAL EDUCATION

In the developing countries, it has been noted that the level of literacy on matters environment as well as the management strategies put in place are quite difficult to measure though there are indicators used in place. As per the study conducted in 1971 which involved the analysis of the existing environmental education programs in the public schools. The outcome of the investigation shows that there are few programs that are operational. In the USA, it was noted that there were few K-12 programs. The positive side was that there were existing programs running in the elementary and secondary schools. It is also noted that the materials produced in matters environments are found to be extensive in amount. They are available in both print and non-print form and have been noted to have an increase in quality.

However, it was noted that in matters interdisciplinary it was found out that there were few environmental concepts that are related to the social sciences or the humanities. In a similar manner, the numbers of the urban oriented materials were found to be few. The schools in urban areas have not yet developed the materials specified to their milieu. In an analysis of the environmental programs that are being developed, it was noted that these programs are developed by single individuals. The same individuals also facilitate the drive for these programs.

The programs found to be successful are those that involved teachers, administrators as well as the citizens of the country. In a bid to promote, environmental education there was the enactment of the law, which is the Environmental Education Act. This act involved funds being set aside to facilitate the enactment of this act. These funds have been set aside to facilitate the projects under environmental education and awareness. For instance, in 1971 the funds were able to facilitate a total of 74 projects while in 1972 the number of projects facilitated was a total of 162.

A major challenge that has been encountered in the act being enforced is the fact that there is a lot of misappropriation of funds. For instance, in the office of environmental education it was noted that a total of 13 million that when projected would have facilitated the realization of over 50 projects. With such funds, being 'lost,' then it causes a pullback in the rate at which individuals are being educated on environmental management. It is a major factor contributing to the large numbers of individuals who lack knowledge on environmental management. This has created a need for drastic measures to be taken in a bid to reduce the rate at which funds are being lost. One of the measures includes the School of Natural Resources (SNR) response (Nath, 2009).

9.9. THE SNR RESPONSE

SNR stands for the school of natural resources. This group is solely responsible for the professionally oriented programs involving the employment in natural resources management as well as environmental education. It also involves the coordinated efforts in all the disciplines in the various colleges involved in the development of environmental programs. This school is guided by an advisory committee representative that is interested in the natural resources as well as the environment. This school is more so applicable in institutions of higher learning. It is among the instructional programs whose major aim is to develop personnel that are capable of handling research, managerial, and educational roles in fields of environment and resources (Nath, 2009).

9.10. ENVIRONMENTAL AWARENESS

In a bid to protect the environment there are certain measures that have been undertaken which include creating awareness on environmental management. It has involved setting up of posters and advertisements that tend to have the exhortation go green. This term though contains a variety of meanings, its major meaning in matters environment us that human beings should be mindful of the natural environment. They do so by making choices that will not impact the environment in a negative way. For instance, this term focuses on people making the right choice instead of purchasing items that have fewer effects on the environment. Take the case where one foregoes purchasing plastic bottles and instead opts to buy bottles made of glass or ceramics (Mishra, 2010).

By so doing they cause a reduction in the number of plastics on the face of the earth. Plastic materials are known to take a long time before they are completely broken down to a percentage that will not be significant. Most of the water bodies such as the oceans have recorded large numbers of plastics that have led to the killing of plants as well as life bodies. Therefore, the go green movement also involves the use of purely natural products instead of those that are made of toxic chemicals. When people participate in the go green movement, they are actually taking part in environmental awareness. The term environmental awareness is defined simply by the words joined to form them. It basically refers to one being aware of the natural environment and being able to make choices that would Benefit the earth instead of harming it (Mishra, 2010).

With the need for change in the environment, environmental awareness has had increased attention among its users. It has involved it being the trend among individuals as it has stirred up debates among individuals. It has less to trends such as environment-friendly homes, people tending towards organic foods and environmental awareness being among the hot topics. There are several areas where environmental awareness can be applied. They include the following:

9.10.1. Application of the Environmental Awareness in Homes

The building materials used have taken a twist as people are leaning towards materials that are more environments friendly. Materials such as asbestos, lead pipes, lead-based paint are known to be harmful to human health and the environment. Communities demonstrate their environmental awareness by deciding to use building materials that are natural and would not cause environmental pollution. Such materials include wood, stone, copper, bricks among others. When using the natural materials, environmental management should be applied so that when the natural resources are used, they are properly used to avoid them being depleted. This will allow the future generation to make use of such resources.

9.10.2. Conservation of Water and Energy

This is among one of the ways in which individuals can demonstrate environmental awareness. Some of the energy sources in use are either renewable or nonrenewable. For the non-renewable sources of energy, care should be taken so that they are properly used. One of the ways in which people can protect their environment includes turning off lights when not in use. In a bid to make use of natural resources available, people have opted to use solar panels to produce the energy they need. There are other alternatives that can be used but scientists are looking into the possibility of solar energy being the main source of power (Mishra, 2010).

The conservation of water is also important. Water is a non-renewable resource and due to its overuse and significant pollution, it has become a limited product. It requires that communities do not waste it. For instance, when brushing one should allow a faucet amount of water running. Water is a very precious and valuable resource that should be properly utilized. Proper management should be put in place to ascertain that this scarce yet vital resource is well maintained in terms of its quantity and quality.

9.11. PURCHASES, RECYCLING, AND ACTIVISM

Rather than purchasing synthetic types of clothing fabric, individuals prefer to purchase those that are made of natural fibers. The reason may be attributed to the fact that recycling has become a major phenomenon known to demonstrate environmental awareness. For individuals who are passionate about it, they have become major environmental activists. They conduct activities such as protests, rallies, or events. Some of the activists have been able to have a presence before dignitaries such as the senate. There they are able to give off the voice for the environment by condemning various acts in the environment such as poaching, extreme mining, and deforestation among others.

They also conduct rallies. The rallies involve them going to remote parts of the country so as to educate the citizens in those places where they can take care and improve their environment. They also conduct activities such as tree planting together with the citizens to also promote environmental management. They champion for reforestation as well as afforestation all in an aim of protecting the environment. They encourage individuals to recycle recyclable products (Mishra, 2010).

9.12. PROMOTING ENVIRONMENTAL AWARENESS AMONG CHILDREN

When it comes to all matters on environmental awareness, it involves participation from the entire society. This means that children will not be left out. Children are given knowledge on the environment so that as they grow, the knowledge is retained and they end up being individuals of substance as they are able to identify the ways in which they can properly manage the environment. As children tend to be creative, they can also come up with ways in which they can help their environment. By impacting this knowledge on children, they are able to have an eco-friendly lifestyle. There are various ways of facilitating this lifestyle among children.

It is important to note that not only is it the role of the parents, but also the society to create environmental awareness among children. A writer once said that it is up to the adults to pass the knowledge to the children and make sure that they start from a tender age adopting their lives in a manner that will take care of the environment even though they are consuming most of its resources (Mishra, 2010). Some of the ways in which one can create environmental awareness among children are discussed in subsections.

9.12.1. Introducing Them to Local Products

This is achieved by supporting local produce. This type of support in production is among the ways in which individuals are able to assist the environment with effects being felt almost immediately. For children to be introduced to such products, it may involve the parents taking them to the local markets. It is among the best ways in which parents are able to explain to their children the sources of food. They tell them where food comes from and that the earth is a crucial resource that nourishes lives. It is important to make them understand that the earth needs tending.

Parents make children to be environmentally conscious by teaching children on purchasing local products, supporting farmers and also eating healthy. It may involve then being in the process of purchases such as allowing them to choose ripe fruits and veggies. They can also participate in the making of healthy foods (Mishra, 2010).

9.12.2. Starting a Garden at Home

A great way in which children can participate in making the environment better is by having them take part in activities that will protect the environment. A good way is planting a vegetable garden with their assistance. After conducting a visit to a farm, the parents may have the children's opinion on what kind of vegetables they can plant in their garden. This may generate their interest in taking part in gardening projects. The parent may start the project on their own and afterwards the children will develop interest and join in creating the garden.

By having, the children take part in such activities it teaches them to be responsible as well as instills discipline in them. It helps them understand that they are capable of bringing change. They should also be engaged in activities such as tree planting and also taking care of domestic animals.

9.12.3. Home Recycling Activities

Recycling is one of the ways people can demonstrate environmental management. It's also one of the ways in which children can be encouraged to be more environments friendly. They are taught how to recycle through use of recycle bins. The parent teaches them proper waste disposal. They will therefore be able to identify what items are recyclable and those that are reusable. One of the ways of teaching such may be through the use of worm farms. They are known to produce worm pee that is helpful in the creation of compost for the home garden (Mishra, 2010).

9.12.4. Proper Use of Both Sides

As children grow, they tend to be active, creative, and have a huge sense of imagination. Most of them take on drawing activities. In a bid to teach them environmental education, instead of them using a blank paper each time they want to draw, they can take the option of recycling the paper. This may involve them drawing on both sides of the page before a blank one is given to them.

9.12.5. Teaching Children About Solar Energy

Solar energy is one of the sources of energy being looked into as a potential major supplier of energy. For a long time, energy conservation has been emphasized. It has involved children being taught to switch off lights as well as plugs for appliances when not in use. Children can be taught to make use of solar energy. For instance, instead of using laundry machines and dishwashers, they can wash their clothes and hang them to dry. This will greatly save on power. For the small children, the parents can tell them stories on the sun portraying it to be something awesome. They can also make use of solar jars which can be used as nights. Generally, it involves the use of solar powered products in the day to day activities.

Children can easily be molded into environment conscious individuals. It's up to the parents to set a good example for them as children tend to emulate the actions of their parents. Parents therefore have the task of being environmental role models. Environmental awareness and environmental education are interconnected. This is because by conducting environmental awareness, one is also carrying out environmental education. They both are therefore processes that allow individuals to explore environmental issues as well as take part in problem solving in the benefit of improving the environment. They both equip individuals with skills so that they are able to make sound decisions on matters environment (Mishra, 2010).

9.13. COMPONENTS OF ENVIRONMENTAL EDUCATION

There are various components of environmental education. They include:

- Awareness and sensitivity. This involves individuals being conscious of the environment as well as the challenges it encounters.
- Attitudes. This deals with the concern for the environment. It also contains the motivation contributing to the improvement and maintenance of the quality of the environment.
- Knowledge and understanding. It involves individuals having information on the environment and the problems faced.
- Skills on how to identify and assist in solving environmental problems.

- Participation in activities that will contribute to solving environmental problems.

9.14. BENEFITS OF EDUCATION AND AWARENESS ON ENVIRONMENTAL MANAGEMENT

There are a number of benefits education and awareness on environmental management. They are discussed in subsections.

9.14.1. Heightened Imagination and Enthusiasm

Environmental education has created an atmosphere that has sparked imagination as well as creativity as it facilitates interactive learning. Once it is integrated into the curriculum, the students become more enthusiastic and have increased engagement in learning. This has been said to increase the achievements in other matters in academics.

9.14.2. Learning Transcends in the Classroom

Besides offering opportunities for experiential learning outside the classroom, environmental education enables the students to make connections as well as apply what they have learnt in real-world scenarios. It also helps them to see the interconnectedness of ecological, social, economic, political, and cultural issues.

9.14.3. Enhances Critical and Creative Thinking Skills

Environmental education encourages students to participate in research and investigations on reasons why and how things happen. They are then able to make decisions on complex environmental issues. As they develop skills such as critical thinking skills. It helps bring up a generation of informed consumers, policymakers, workers or decision-makers.

9.14.4. It Supports Tolerance and Understanding

It encourages students to conduct investigations on varying sides of issues in order to understand the full picture. In the long run, it promotes tolerance of the varying points of view as well as the cultures.

9.14.5. It Allows State and National Learning Standards to Be Met for the Multiple Subjects

When environmental education is incorporated into the curriculum, it allows teachers to integrate subjects such as science, math, language arts, and history among others to be a richer activity or lesson. It therefore satisfies multiple state and national academic standards in the various subject areas.

9.14.6. Declination in Biphobia and Nature Deficit Disorder

When the students are exposed to nature and they are allowed to learn and play outside, the environmental education creates a sense of sensitivity, appreciation, and respect for the environment. It therefore combats the nature deficit disorder.

9.14.7. It Encourages Healthy Lifestyle

Environmental education allows students to be active and outside. They will be able to address some of the health issues evident in children such as obesity, deficit disorder, attention, and depression. It allows lifestyle aspects such as good nutrition can be emphasized. It is also one of the ways in which stress can be reduced as much of the time is spent in nature.

9.14.8. Strengthening of the Community

It creates a sense of places and connection. This is achieved through community involvement. It allows people to reach out to community donors, experts, volunteers, and local facilities in an aim of bringing the community together. This will create a sense of understanding and creates a platform where environmental issues can be discussed.

9.14.9. It Fosters Responsibility for a Better Environment

Through environmental education and awareness, people become more aware and sensitive to the environment. They are able to make better decisions that will lead to the improvement in the status of the environment.

9.15. CHALLENGES IN ENVIRONMENTAL EDUCATION

There are various challenges encountered by teachers in implementing environmental education. The major ones being time constraints as well as the work overload placed on the teachers. Some of the time barriers include teachers not having much time to bring out the students from the class to conduct all the natural activities. They have therefore been limited to telling the students stories on such. The teacher is tasked with the duty of conducting research on these topics so that when they give the stories, they are able to make it as real as possible.

There is also the challenge that most of these teachers may not have the adequate training on environmental education and issues pertaining to it. It brings the problem that the teachers are unable to make adequate planning on these issues and are unfit to integrate environmental education as a subject matter in teaching. The other challenge is that there is no course that is specified to environmental education. Therefore, the teacher has to plan for the outdoor education which may involve them asking students to identify certain plants in the environment. In the long run the students are able to learn more about their environment (Mishra, 2010).

It was also noted that most of the teachers lacked knowledge on the handbook on Environmental Education on the curriculum that is provided by the Ministry of Education. Thereby proving that efforts made by the government in facilitating environmental education are futile. A study shows that most of the schools lack copies of this handbook. Some of the teachers have even admitted that they were not aware that the handbook was given off by the ministry of education. This handbook is crucial as it is one of the tools that teachers use to conduct Environmental Education. It is a major source of information on matters environment.

Though some of this information is available on the internet, the teachers are not able to access it as most of these schools lack technological tools to facilitate this. Most of them even lack classrooms therefore, the access to these tools is limited. Some of the schools are able to access the internet but they face a problem with the internet connection. Some of the internet services provided is not efficient. There is also the problem of lack of teaching aids that could be used to facilitate environmental education. Some of the teachers also lack information on matters environment. Such teachers end up being rendered unable to offer environmental education (Mishra, 2010).

There is also the challenge of misappropriation of funds. This applies in the case of organizations created to make projects that will contribute to the increase in education and awareness on environmental education. Most of these organizations end up failing due to managerial problems. When funds given to these companies are lost then they will be unable to achieve and realize the set out projects. Therefore, the rate of increase in education and awareness in environmental education will reduce. These are but some of the problems encountered (Mishra, 2010).

10

Gender and Environmental Management

CONTENTS

10.1. INTRODUCTION

Equality of the sexes is defined as a state of equal access_N to resources and opportunities and basic rights, disregarding gender. It is considered as one of the most pressing unfinished businesses of the modern era. Gender equality has become a movement much bigger than it is strongly tied to women's rights as very many practices even culturally from a historical point of view were and still are, harmful to the female gender. Gender equality was never widely advocated until the post-war era but there are some notable proponents that could be mentioned with one of the earliest being Christine de Pizan who was a poet and writer in the King's Court during the reign of Charles the 6th of France in the 13th century. Her most remarkable work was a lengthy prose written in vernacular French called "The Book of the City of Ladies" which preached on irrational Prejudice being the main reason for the oppression of women and furthermore highlighted but numerous advances in Society would not have come without women. She discussed powerful and influential women like the Queen of Sheba, Mary Magdalene, Blanche of France, and many more whose merits led to great accomplishments.

The 16th century also saw the creation of an evangelical group known as the Shakers or Quakers who's main preaching Focus was on the belief in the ability of each person accessing the "light within" or "that of God in everyone." This is led by Jones Michelle who had a revelation that all sexes should be equal and this system of gender balance based on spiritual welfare is a behavior of gender-balanced lasting over 200 years. Many may conclude that they were a radical religious sect but they made real the practice of gender equality and made it something quantifiable and achievable (Buckingham, 2005).

Despite the aforementioned, the suffrage movement is what is most commonly recognized as the real tangible beginning of movements towards gender equality. This is mainly because it happened on a wider scope in relation to society and western cultures primarily. It became a militant movement that sought to allow women the right to vote and hold public office and lead to more significant changes to women's property rights especially in relation to marital status.

Women's rights and many of the movements associated with it were a post-war era phenomenon with the United Nations and many other international agencies promoting gender equality by adopting various conventions. Many expert sociologists attribute this massive surge in gender equality movements to the Invention of the contraceptive pill in the 50s that allowed women to focus on more than just childbearing and taking care of children. Whether or not this is factual not the question to be answered here however, it may be a factor in itself. What can be agreed, however, is that legislation and policies that had strong affirmative action were the critical turnover points in bringing societal changes and changes in attitudes. A few important legal changes, conventions, and protocols made include:

- The convention on the elimination of all forms of discrimination against women which is an international Bill of Rights for women (1979).
- The Beijing declaration where the UN adopted a resolution during the fourth world conference for women that promulgated principles that are set on gender equality concerns (1995).
- The Maputo protocol that was adopted by the African Union guaranteeing rights to women especially in controlling reproductive health, female genital mutilation, and many African cultural practices that threatened women's rights.

Many other legislative changes made to combat violence against women and domestic violence, promote equality in the workplace and end discrimination against women were made but the changing of social attitudes have made a bigger impression still in the direction towards gender equality. This includes the changing of a woman taking her husband's surname on marriage becoming a more non-automatic trend. Other social manifestations include the increase of men in occupations previously considered women's work like nursing and cleaning or the role of child-rearing or child-rearing becoming less exclusively female-oriented. These practices and philosophies are what are commonly referred to as gender neutrality and gender equity helping achieve the end goal of gender equality (Buckingham, 2005).

The Social and ideological backdrop of gender equality has its flaws in today's world through certain practices that are still common like the cultural preference for a male child in China, for example, or the failure for countries to adopt policies quickly like the United States having a worse maternity leave than Costa Rica. The encouraging news is that the United Nations and other international bodies have visions clearly stated for women being treated equally on social, economic, political platforms and all other aspects of society even in the global south.

The achievements of gender equality require gender-responsive institutions and robust governance and accountability systems noting that social and cultural restrictions, especially from gender stereotypes, are limiting. For real gender equality to exist the entire population most actively supports equal rights and opportunities (Buckingham, 2005).

10.2. A BRIEF HISTORY OF GENDER RELATIONS ENVIRONMENTAL CONSERVATION

In 1985, one of the first international forums that fostered the relationship between sustainable development and women's involvement and empowerment was held in Nairobi. The environment was listed as one of the areas of concern for women. As much as I did not directly make any recommendations, the document delineated sustainable development and women bringing forward debates from the standpoint. This was a step forward as the United Nations Conference on the Human Environment held in Stockholm 1972 did not even mention women and the term man was used in a general sense. This has been an issue recognized since the outcomes of the gender plan of actions 2006–2011. From a programmatic level, there has been good effort made in integrating gender language into project documents to make reference to women's gender equality.

Gender equality is a goal in itself but is paramount in making progress to the achievement of general development goals, more importantly environmental sustainability. You can unanimously state that women bear the highest cost of environmental problems but have the greatest potential for contributing to solving the crisis. This was brought to the forefront in the 1992 Rio de Janeiro United Nations conference environment and development (UNCED).

Women stand to gain the most beneficial impacts from environmental conservation by the context specific solutions that they are able to develop especially in households and communities. In 1984, a massive step forward was made by the United Nations Environment Program UNEP as they built the senior women's advisory group on sustainable development which held several studies on the connection between the exclusion of women and the roles they play in terms of environmental development. One great example of the successful involvement of women promoting the participation by and training of women as providers of water in the international drinking water supply and sanitation decade (1980 to 1990) and international conference on water and environment that was held in Dublin in 1992.

Work continued to culminate with the Global Assembly on Women and the Environment: Partners in Life where presentations showing 218 successful experiments that were carried out by women in developing countries highlighted the importance of women involvement and validated recommendations to the United Nations Conference. This is the origin of Agenda 21 that is aimed to initiate a set of sustainable development action plans by the United Nations.

10.2.1. Agenda 21

Agenda 21 was known as 2021 was the original target here in which Development Goals were meant to be achieved. It is a 351-page document with 40 chapters and more than 145 references specifically mentioning the rules women play in environmental and sustainable development. Chapter 24 is titled global action for women towards sustainable development and it acknowledges the importance of the participation of women. Section 3 of the 4 sections of the document is on the strengthening of the role of major groups and that is inclusive of children, youth, and women. Among the proposals made as action items against discrimination include:

- Implementation of strong institutions non-governmental organizations and women's group in general that can train and use management resources towards environmental sustainability.
- Initiatives to reduce the Enormous workload and various unequal circumstances borne by women e.g., division of household tasks.
- Application and setting up of programs to promote preventive and curative health services for women and voluntary family planning services that are safe and inexpensive.

178 governments pledged their vote to adopt the program after much negotiation and consultation. Of course, implementation by any of these governments is voluntary and the fashion of adoption varies widely but the first most important step of discussion in 1992 culminated in many more conferences. With the identification of this specific subject matter, donor agencies open their doors and international organizations invested in the aspects of gender environment relationships and activism (Buckingham, 2005).

This became more evident with the declaration on social development produced by the world summit for social development in Copenhagen 1995 that stated "We acknowledge that social and economic development cannot be secured in a sustainable way without the full participation of women and that equality and equity between women and men is a priority for the international community and as such must be at the center of economic and social development. Headway was being made but application was still in need; countries still treated women as a "vulnerable group" not reflecting well on the genuine incorporation of genders.

This was highlighted in 1997 in the fifth UNCED conference that revealed the word deficiencies in the gender-based approach and this was attributed to changes in Economic and political structures not having often enough impact to overcome other developmental issues like poverty. Again the United Nations Commission on the status for when highlighting the importance of the gender-based approach but focused on securing greater independence for women in order to promote their involvement in sustainable development. The Millennium Summit in New York set up millennium development goals (MDGs) includes MGD 3 centers on Gender equality and MGD 7 set on environmental sustainability continuing to build on both as important development agendas in the post-2015 framework (Buckingham, 2005).

This brief history can be concluded by noting that both separate but intertwined issues stemmed from prominent movements. We should give thanks to feminist movements and environmental movements and not make the initiative compete with one another but rather combine them in a complementary philosophy. Gender mainstreaming would be a monumental addition, identifying the interests of men and women and their experience in the development process. A great achievement would be to entirely demystify gender mainstreaming by unequivocally defining it and highlighting the practicality of addressing gender inequalities and how that will further environmental management interventions.

10.3. STATISTICS ON GENDER AND ENVIRONMENT

As already stated above gender inequality exists but these gender inequalities affect environmental issues and planning effectively around some of the already acquired norms would be effective towards successful environmental development. Some statistics to keep in mind or listed below.

A study by the International Union for the Conservation of Nature (IUCN) on disaster and gender statistics show that gender differentiation carries when dealing with natural disasters. Women are more likely to die than men in natural disasters and the aftermath of natural disasters impact women more than men (this is in relation to food shortages and economic resources). It is important to note however, that when women hold higher socioeconomic statuses the increased gender equality creates more resilient communities (Arora-Jonsson, 2013).

- **Takeaway:** A gender-neutral society is more efficiently equipped disaster management.

According to the World Health Organization, 3 billion people cook using polluting fuels and 4 million people die illnesses attributed to these pollutants. A disproportionate number of exposed are women.

- **Takeaway:** Household chores and many similar activities are still culturally highly gender influenced.

Children are 60–85% more likely to suffer from malnutrition in areas where land rights and ownership have a higher gender inequality according to the UNDP, UNPBSO UNEP, and the United Nations Entity for Gender Equality and Empowerment of Women.

- **Takeaway:** Women are great custodians of land and influential custodians of resources. Gender inequality promote a monopolistic mentality where the needs of the few (men) outweigh the needs of the many (women and children)

There was a 10% increase from 2015 to 2017 in female representation in ministerial positions to do with the environment globally.

- **Takeaway:** Unequal representation of women in politics translates to less reflective and therefore less effective governance on environmental matters and meeting environmental needs of the population.

These facts and figures mentioned have but a few of the mini crippling faults that gender inequality bears along the political and socio-economic spectrum. Other economic issues include the lack of support of women in the green economy. I'm there for the lack of Engagement informal economic activities that negate the creation of new opportunities and overall development.

The above-mentioned statistics and more should reinforce the value of a gender-inclusive approach where we incorporate a gender perspective across all spectrums, inclusive of environmental development and sustainability. To have more effective sustainable projects, men, and women should be integrated in different positions and different stages of environmental project management. This should be done by developing an understanding of the different roles, responsibilities, norms, and priorities between men and women shaping human interaction in the environment (essentially mainstreaming gender in all programs, projects, and activities).

10.4. GENDER MAINSTREAMING

Gender mainstreaming is a policy practice and concept of assessing the various implications for four different genders in any planned action including legislation programs policies in all areas and all levels. The 1985 world conference of women in Nairobi is where the concept was first proposed and it was an idea to push United Nations community development, however, it was made formal 10 years later in the fourth world conference on women in Beijing China (MacGregor, 2017).

Its main principles are to prioritize gender equality in a holistic approach by assessing every policy or piece of legislation on whether they promote or diminish gender inequalities. The incorporation of gender into politics and decision-making is also another key point as historically international and peacekeeping agreements and generally most legal resolutions Project the same stereotypes that disempower women. This is particularly important in conflict resolution and peace-building where statistically women suffer more than men. Generally, there needs to be a whole shift in culture especially in institutions and organizations that have in the past perpetuating gender inequality. There are many places where gender mainstreaming has brought a paradigm shift in the whole mechanisms that run a nation. A great example is in Nicaragua

In 1990, Nicaragua had its first female president Violetta Chammoro who revitalized the outdated Sandinista women's organization. Under the Chammoro government, it was renamed the Nicaraguan Institute for women and encouraged the involvement of Nicaraguan women in the country's socioeconomic cultural and political development bringing changes in traditional gender constructs. Although the United Nations general assembly noticed some flaws such as the backlog of women 's rights legislation, the institution was instrumental in bringing changes that promoted gender equality in agriculture and socio-economic developments (MacGregor, 2017).

A more successful example would be equality policies implemented in the European Union which initially was founded on three pillars anti-discrimination affirmative action policies and gender mainstreaming. Several governance initiatives in the European Union developed and examples are the European employment strategy that encourages governments to adopt compliance measures that are lined with gender mainstreaming when deciding on employment policy. The transnational women's network that worked on the EU stage to encourage the coordination of women's Civic groups and European Institute for gender equality that mandated to "provide expertise improve knowledge and raise visibility of equality between men and women" are both of you of many governance actors on the European platform boosting the idea of gender mainstreaming.

It's important to know that experience overall has been mixed as even though the philosophy has been widely accepted there has been a weak level of commitment on improving women's quality of work and life overall. Poor implementation hindering progress and ineffective results has been some of the many criticisms offered by the United Nations in the Promotion of gender mainstreaming (MacGregor, 2017).

On speaking on gender mainstreaming in environmental management it is crucial to take in account the different needs, roles, and expectations of women and men in sustainable development. Gender-neutral projects, on the other hand, assume that the needs of men and women are the same and so are their priorities opportunities and expectations and these risks reinforcing already existing gender-based discrimination. Environmental management project should be sensitive to gender by raising several questions:

- Are women and men's experience taking into account sustainable development management of resources to achieve optimum results?
- Is it recognized that effect of various hazards like toxic chemicals are different in men and women?
- Are the different levels of knowledge and skills of men and women considered and used appropriately in development?
- Is attention given to foster participation in both men and women and is opportunity given in the decision-making process and leadership roles?

Men tend to have a bigger ecological footprint than women and the production and consumption patterns prove that women are more resource-efficient, more likely to recycle, and have wide knowledge about their local environment. This, however, does not balance out in the sense that men still occupy most green jobs and have better access to green technologies.

Due the fact that equal opportunities of men and women go further than just creating a more equal Society but also generates new needs and encourages new development of product, it means that it is much more than just a women's issue and has the advantage of stimulating economic growth creating higher-level jobs, supporting communities, reducing poverty and raising productivity (MacGregor, 2017) (Figure 10.1).

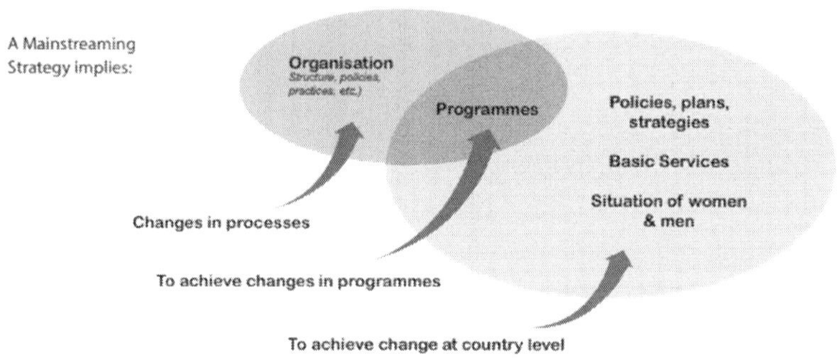

Figure 10.1: The three spheres of gender mainstreaming.

Source: https://wedocs.unep.org/bitstream/handle/20.500.11822/7655/-Gender_equality_and_the_environment_Policy_and_strategy-2015Gender_equality_and_the_environment_policy_and_strategy.pdf.pdf?sequence=3&isAllowed=y.

10.4.1. Modern Examples of Gender Mainstreaming (UNEP and OSCE)

10.4.1.1. Gender Mainstreaming in the UN

Gender mainstreaming has been adopted into the UN business model and it emphasizes two areas in which gender mainstreaming has to occur:

- The level of the organization or institution; and
- The level of the program.

Require specific governance and advisory structures to be established in order to assist the implementation of gender mainstreaming. We will take a look at two case studies the first being the United Nations environmental programs (UNEPs) gender architecture that compromise is a steering board a committee in charge of implementation and cross-organizational teams for thematic and technical gender using gender focal teams (MacGregor, 2017) (Figure 10.2).

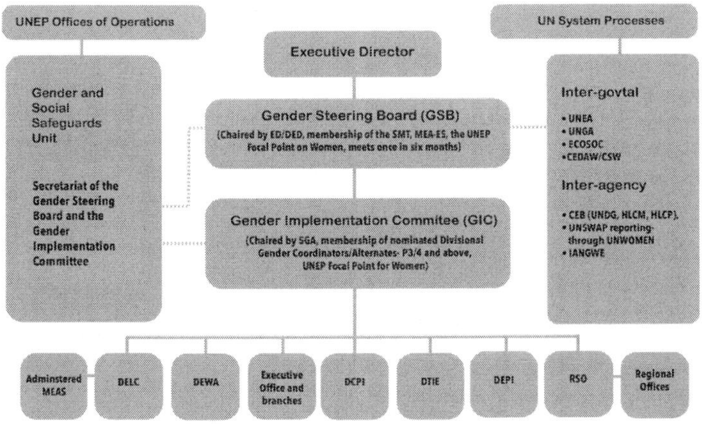

Figure 10.2: UNEPS gender architecture.

Source: https://wedocs.unep.org/bitstream/handle/20.500.11822/7655/-Gender_equality_and_the_environment_Policy_and_strategy-2015Gender_equality_and_the_environment_policy_and_strategy.pdf.pdf?sequence=3&isAllowed=y.

10.4.1.3. Rules and Responsibilities Within the Gender Architecture of UNEP

- **The Gender Steering Board:** The executive director has the ultimate responsibility for treatment of results and expresses his goals through a contract with the secretary-general and overseeing the senior management team. Under the senior management team is an agenda steering board that meets at least twice a year, headed by the executive director or deputy executive director to ascertain implementation of programs while explicitly assessing the level of gender mainstreaming.
- **Gender Coordinators:** Is division director is meant to a point a staff member to take up the role of gender coordinator for The Division that ensures necessary gender mainstreaming
- **Gender and Social Safeguard3 Unit:** The DSS you are responsible for the coordination monitoring management and

capacity building of gender mainstreaming across the entire organization. They work within the office of operations and report to the deputy executive director or the chief of operations. Therefore, the responsible for overseeing the integration of Environmental development with social-economic and gender safeguards

- **Focal Point for Women:** The focal point for women is an appointed key member for the gender architecture and has access to all information necessary for the functioning of The Office and is also located in the office of operations, working closely with the senior gender advisor.

Categorically all members and general staff have complimentary responsibilities in the aim of gender mainstreaming with special focus on financial and human resources. This is the way the UNEP organizers self however, many other organizations have taken the concept of gender mainstreaming and adopted it to suit their own institutions. Another briefcase study that can be observed on the implementation of gender mainstreaming is with the organization for security and cooperation in the European Office of Child Support Enforcement (OSCE).

10.4.1.3. Gender Mainstreaming in OSCE

The organization for security and cooperation in Europe is the world's largest security-oriented inter-government organization that promotes practices in the Promotion of Human Rights more especially women's rights along with many other mandated issues. They realize the relationship between human society and the environment is not gender-neutral upon close examination. Men and women have different personal attitudes, priorities, and power over resources in relation to Environmental Protection and environmental degradation exacerbates the intensifying conflict, and women and men must take up the different responsibilities (Dankelman, 2012). They insist on gender mainstreaming in the project cycle pain explicit attention on each of the four steps are shown below (Figure 10.3):

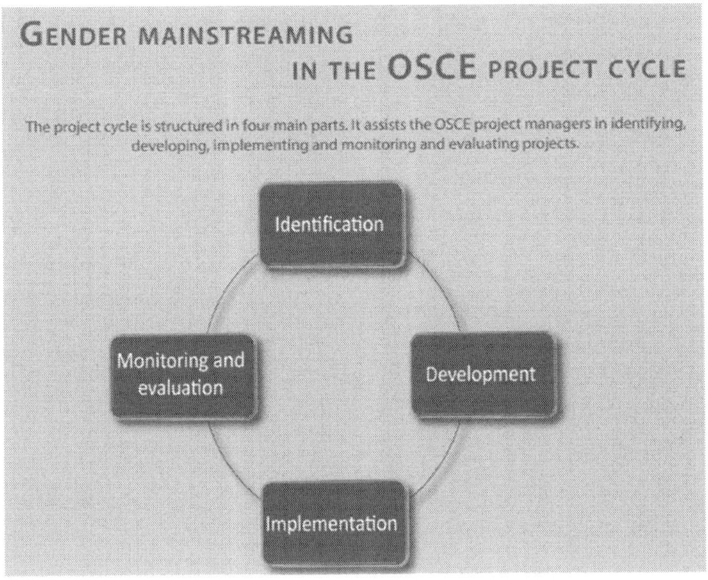

Figure 10.3: Gender mainstreaming in the OSCE project cycle.

Source: https://www.osce.org/gender/36360?download=true.

- **Stage One: Product Identification:** The stage includes ascertaining a product vision, breaking down of a situation analysis and assessment of the participation analysis. These areas to analyses include:
- **Policy Framework:** What are the national obligations and commitments to gender equality actively followed.
- **Institutional Structure:** What is the proportion of women and women in environment-related decision-making positions?
- **Sectoral Analysis:** What roles do men and women play in different sectors, for example, forestry agriculture fishery water lands, etc.
- **Existing National Policies:** How do they extend the benefits and opportunities available to women and men?
- **Legal Frameworks:** What are the laws on ownership access and use of natural resources and do they promote gender equality.
- **Budgetary Allocations:** How does budget allocation impact men and women in the environmental field.

- **Environment and Health Indicators:** What differences are shown in the impact of pollution on men and women?
- **Gender Stereotypes and Cultural Norms:** In which fashion are gender roles enforced and how does that affect participation in project management.
- **Role and Men and Women Regarding Energy and Natural Resources:** How much time is spent on activities relating to Environmental management and what is the relative income of women and men in these sectors.
- **Media and Public Perception:** How well do the media cover topics related to environment and gender and the nature of this coverage?
- **NGOs and Other Institutions:** Are there any organizations working on environmental and gender issues and what is the niche and what are their successes and failures?
- **Participation Levels:** Do men and women share equal access to participation and what contributes or serve as a hindrance to their participation.
- **Stage 2: Project Developments:** This stage comes with the development of the project proposal that outlines the objectives, activities, and what should be the expected results. The project objectives should be gender-sensitive. At this level, questions should be asked in relation to developing specific project objectives relating to gender and specifically what activities are required to ensure attention is paid to gender issues. Management arrangements should also be designed to be gender-sensitive where we ascertain that all project personnel are familiar with the gender issues and how they relate to the project. From the managerial decisions are made on the composition of the project personnel and whether training on gender and environment issues is required only within their own team but on potential partners.
- **Stage 3: Project Implementation:** As the project develops, a keen eye should be used to identify any issues that bring an imbalance on gender issues. The report's Brief and speeches should be given in a gender-sensitive language and unless the product is focusing on specific reasons on only men and women the gender mainstreaming should be ensured in all project activities. Resources are also tracked along the same narrative together with the implications the project may have.

- **Stage 4: Project Evaluation:** The stage of monitoring and evaluation is to assess the degree in which the project produced intended results and with the use of gender, mainstreaming more specifically in regards to gender issues. The evaluation report should show that there was a clear identification of gender issues with respect to the different needs and priorities of men and women as well as what input they both have. Recommendations and lessons learned on gender perspectives should be disseminated as well.

The above techniques employed by the United Nations and Organization for Security and Cooperation in Europe for gender mainstreaming are but a few of the techniques that an institution can use to promote gender equality. The important parts for the organization is being able to recognize the differences in gender and promoting gender sensitivity through training and practice (Dankelman, 2012).

10.5. GENDER ISSUES AND CLIMATE CHANGE

Climate change has impacts across geographical regions but also across the gender spectrum. This is because gender inequality is translated into economic, political, and social inequality. Although the overall objective would be to move towards climate-resilient and low emission strategies for sustainable development, the design must have specific gender components in outcome outputs and activities (Figure 10.4).

Figure 10.4: Both women and men can be agents of change in all the different levels of the process by increasing access to information and services forming gender-sensitive adaptation plans.

Source: https://electricityhr.ca/news-events/signature-events/agents-of-change/.

Mitigation efforts in climate change things that both men and women should be supported as entrepreneurs towards renewable energy low-carbon transport and climate-smart agriculture, for example. Clean energy can contribute significantly to the well-being of women, men, and children alike through such efforts like the production marketing and sales of solar panels and energy-efficient cooking stoves. This would automatically solve the problem of health pollutants being a risk for women in the household. It would also alleviate some of the domestic workload and the risk of dangers faced when collecting solid fuels like firewood. (Risks include domestic violence) (Dankelman, 2012).

This becomes particularly more important in rural societies in developing countries that predominantly rely on natural resources for energy black charcoal and agricultural waste wood and mangrove. This is a burden disproportionately borne by women. The 7th sustainable Development Goal states that women can be powerful agents of change in the transition towards renewable energy especially with technology which the prices are rapidly falling. The more impactful change, however, would be through raising awareness and promoting gender policies in institutions and organizations that are dominated by men. This involves changing the gender trend on issues like Investments tariffs pricing and participation.

Transportation is another sector in climate change where appropriate initiatives must be undertaken specially to promote low-carbon and non-motorized efforts. This must take into the consideration the different opportunities and risks for both women and men. In Kenya, for example, women face sexual abuse and harassment while simply waiting to use public transport. It is a rampant issue but it's very rarely reported.

In the Global North gender, issues in the transportation sector differ however, there is still a disparity in relation to efforts made towards climate change. The use of public transportation walking and cycling is much more common among women than men and so it makes a lot of sense to engage both women and men in urban transportation planning and development (Dankelman, 2012).

One of the biggest contributors to climate change is deforestation and forest degradation. Gender mainstreaming in efforts to curb deforestation and forest degradation has been minimal because women's involvement in the sustainable development plans have been superficial at best in relation to men and to exacerbate the problem women are not as informed about the related activities as their male counterparts (Buckingham, 2005).

The United Nations has made efforts in ensuring program offices are gender-sensitive in the planning implementation and monitoring of climate change projects. This involves but is not entirely limited to usage of analytical framework and tools to integrate gender-based considerations and also supporting the connection allowances and use of sex disjointed data in various climate-sensitive sectors.

10.6. GENDER ISSUES IN NATURAL DISASTERS AND CONFLICTS

Gender and environmental management cannot be discussed fully without covering the gender issues involved in natural disasters and conflicts. There are distinct and very different challenges faced by men and women in disasters and violent conflicts and consequently the roles and responsibilities in mitigating these issues also differ.

There are gender differences in the environmental impacts of disasters and conflicts and they do not favor women. They suffer specific risks from the exposure doing the gender rules, for example, sexual violence while performing daily tasks like collecting natural resources like fuelwood and water. Emergency responses and humanitarian settings during these natural disasters and Conflict should be geared towards gender differences. Projects should be integrated with the gender perspective in the design and implementation of risk management. This can be done by studies that analyze how environmental damage from disasters and conflicts disproportionately affected groups like women and children and incorporate gender-specific analytical tools for these data collection methods. Again, awareness of these issues is the first step and more efficient and objective data analysis will help us determine how to move forward on a gender mainstreamed platform. This will help support and enhance the role and responsibility of women as active actors in post-crisis settings (Buckingham, 2005).

There are opportunities in post-disaster and post-conflict situation to achieving a more inclusive, equitable, and sustainable recovery like:

- Political empowerment by allowing thc management of the environment and natural resources
- Economic empowerment by the creation of sustainable income generation

A successful project example was the conflict and dictator risk reduction in Lucaya basin in the Democratic Republic of Congo. Here the UNEP brought together different stakeholders to support work towards shared development in the area that was fostered by women being active community leaders, farmers, and income earners engaged throughout the project. The women in local leadership played a tremendous role in the ecosystem disaster risk reduction activities and managing priorities for water and land. This success was influenced by the nature of the project being community-based and the effort driven by the heavy involvement of women.

10.7. GENDER AND ECOSYSTEM MANAGEMENT

Healthy and productive ecosystems are a priority area in environmental management with gender inequalities being an important factor, shaping the use, management, and conservation of biological resources. Biodiversity and the conservation of it are important for general human well-being for the present and future generations. Sex-disaggregated data highlight many key gender issues that influence biodiversity loss, conservation, and sustainable use. One of the most concerning is that in many parts of the world women enjoy less access to and less control of biological resources particularly in regards to land. This is particularly worrying because they are primary resource users and therefore should play the role of managers more actively in safeguarding ecosystems' health and functioning but the inclusion of gender indicators in biodiversity are lacking. Ecosystem economics is also another area where gender dimensions should be paid close attention to because of the fact that women have been underrepresented and the deeper reason for this is education for sustainability has lacked gender considerations (Arora-Jonsson, 2013).

A notable effort towards this specific cause was made by the UNEP in 2016 through a pilot program that was conducted to increase the capacity in the developing world integrate gender into the national biodiversity strategies and action plans (NBSAPs) that stimulated the involved countries to mainstream gender into their work under the convention on biological diversity (Arora-Jonsson, 2013).

10.8. GENDER AND ENVIRONMENTAL GOVERNANCE

Environmental governance is a set of concepts based on political ecology and environmental policy advocating for sustainability and managing human activities. It is these key ideas that help conservation managers and scientists to participate effectively and efficiently. Due to the fact that gender and basic governance and politics have shown distress in gender, mainstreaming it is difficult to imagine that the subset of environmental governance would. In fact, less than 10% of UN member states finance ministers are women giving you an idea on how gender bias the green economy is. The overall unequal presentation of women in political positions is an added factor showing environmental governance to be very inefficient. The insights of both men and women in financing environmental and Climate Change initiatives together with equal participation in decision-making foster effective management (Arora-Jonsson, 2013).

When it comes to the protection of biodiversity and governance in the environmental sector illegal, wildlife trade is another area where perceptions of gender have an effect. The marketing and selling of illegal wildlife has a huge input from gender perception in that many are culturally valued for their claimed medicinal properties (e.g., creasing virility, masculinity, or fertility).

10.9. GENDER RELATIONS IN CHEMICALS, WASTE AND AIR QUALITY

Many environmental management systems (EMS) have been aimed to promote the effective management of chemicals and waste to minimize the impacts on the environment and well-being of human beings. It is important to state that men and women are exposed to different chemicals, wastes, and even air qualities and therefore suffer different health impacts. Three key groups of chemical substances are listed as being of concern and they are persistent bioaccumulative and toxic and include persistent organic pollutants, heavy metals, and endocrine-disrupting chemicals. The different physiology of men and women and predispositions of both also play a factor in the gender-related aspect of chemical management. The strategic entry points for gender mainstreaming would be to monitor and review trends in chemical production handling movement news release and disposal together with the environmental health and social-economic impact while simultaneously employing a gender disjointed approach.

10.10. GENDER AND RESOURCE EFFICIENCY

Gender mainstreaming towards resource efficiency means that one must develop smart ideas and messages to appeal to both men and women to promote consumption of resources responsibly but even before that promoting equal opportunities for women and men in the green economy. Gender is a key component when formulating policies and programs related to resource efficiency and the best way is to follow a gender-responsive approach that considers different needs roles and responsibilities for both genders (Arora-Jonsson, 2013).

The UNEP has had several projects that serve as examples on correct resource efficiency with gender considerations, one of them being a project aimed to transition Pakistan into an energy-efficient country that looked at both men and women as important actors in the energy sector as well as beneficiaries. Although the work mainly consisted of literary reviews and Academic work gender consultation was highly involved in a project like this if it wasn't existing inequalities would have been enforced or exacerbated (Arora-Jonsson, 2013).

10.11. ECOFEMINISM

The relationship of the relationship of gender and environmental management cannot be fully explicated without involving ecofeminism (Figure 10.5).

Figure 10.5: Ecofeminism is social political movement running along the lines of feminism that is environmentalism and the overall relationship between women and the earth as the fundamental analysis and practice.

Source: https://sustyvibes.com/the-meaning-of-ecofeminism/.

Ecofeminism draws on the same aforementioned concept of gender into the relationship of humans and the Natural World.

Traditional literature like Francoise de Eaubonne Le Feminism ou La Mort (1974), Women in Nature by Susan Griffin and the Death of Nature by Carol Merchant are texts that propel the association between oppression and domination of women and generally all marginalized groups to the oppression and domination of nature. Gender in nature was appropriated by these kinds of influential texts expressing that capitalism and other existing patterns in society only propagate values that do not benefit women. Some of the concepts that ecofeminism appropriates include:

- Modern science in ecofeminism where a paradigm shift is being created from a theme of modern science being a projection of Western men values to where scientific knowledge is being accepted as a universal value-free system.
- Vegetarian ecofeminism that applies ecofeminism to animal rights by expressing the oppression of animals from a feminist analysis. It shows me eating as a form of Patrick domination suggests linkages between male aggression and violence with meat-based diets.
- Materialistic feminism is another common dimension that looks to analyze the connection between various male-dominated institutions such as labor power property as an expression of male domination over women and nature. From this angle, ecofeminism takes to eliminate social hierarchies all of which favor and are dominated by men over biological and social reproduction

10.11.1. Spiritual Ecofeminism and Cultural Ecofeminism

Spiritual ecofeminism focuses on the values of compassion and non-violence preaching on an earth-based spirituality that recognizes the earth as a living entity and human beings within an interconnected community. Ancient traditions like the worship of Gaia and Mother Nature are some of the common practices and this kind of ecofeminism is not linked to one particular religion.

10.11.2. Ecofeminism Movements

Movements of the 1970s and 1980s saw women working efforts to protect wildlife food and water participating in various environmental movements.

In Uttarakhand India during the early 1970s, women mobilize and took part in the Chipko movement that was series of nonviolent protest against logging to protect forests from deforestation. This vast area along the foothills of the Himalayas successfully protected and the world's immediately took notice. In reality, many of its leaders were both men and women who practice methods of Satyagraha or truth holding and nonviolent movements that was inspiration to many future initiatives. The Chipko movement was awarded the right livelihood award which is an international award honoring and supporting those offering practical and example answers to the world's most urgent challenges. They were recognized for their dedication to the conservation-restoration and ecologically sound use of India's natural resources (MacGregor, 2017).

The Green Belt Movement in 1977 was initiated by political activist and Nobel Peace Prize Winner Professor Wangari Maathai. She was known as a protector of human rights and the lover of trees speaking "What a friend we have in a tree, the tree is a symbol of hope, self-improvement, and what people can do for themselves." The Green Belt Movement is an indigenous grassroots NGO from Nairobi Kenya that mobilized local communities and their consciousness towards justice equity reduction in poverty and most importantly environmental conservation using tree planting as a main focus. Since the Foundation of the movement in 1977, over 15 million trees have been planted. The foundation has a direct political context in relation to women involved. In the beginning women involvement, included tree planting programs but it's slowly grown beyond that as the green belt movement started providing free lessons to women in rural areas to grow food and feed their families. This hands-on intensive approach improved many communities through awareness and access to two natural resources. The unfortunate reality though was that the highly patriarchal Kenyan system continued to prevent women from accessing equal opportunities on the social-economic and political spectrum and prevented women from being involved in the decision-making process (MacGregor, 2017).

The love canal mentioned earlier toxic dump site in New York around the Niagara River also bore the honeycomb feminist movement through mother and environmentalist Lois Gibbs. She led her community in protest after discovering the entire neighborhood was built on top of a toxic dump site that's caused illnesses among children and reproductive issues among women. Over 750 families were evacuated from the area and relocated by the Federal Government because of efforts made in the movement.

The women's pentagon action of the 1980s is another example of an ecofeminism movement that combined the philosophies of feminism and ecology linked to the patriarchal oppression of women and the Earth. They organized a peaceful protest at the Pentagon where women stood hand-in-hand against ecological destruction and nuclear weapons. The Express feminist critique on the nuclear Arms Race during the Cold War and they are starting today as one of the bridges between the Reagan era American feminism and the modern-day Environmental activism (MacGregor, 2017).

The Akwesasne mother's Milk Project is a great example of gender mainstreaming and government-funded research. It was launched by Katsi Cook who is a Mohawk Native American midwife and environmentalist. Her Mohawk reservations water supply was investigated and it was concluded that high-levels of contaminants in the water impacted their babies. Their children were exposed to 200% more toxins than children not on the reservation through their mother's breast milk. This translated to a high number of miscarriages, birth defects, and many other health issues. The mother's Milk monitoring Project and the women's dance health program are better few of services bounded by Katsi Cook that function to this day. Cook eloquently stated that "Women are the first environment. We are an embodiment of our mother Earth. From the bodies of women flows the relationship of the Generations both to society and the Natural World. With our bodies, we nourish sustain and create connected relationships and interdependence. In this way, the Earth is our mother, our ancestors said. In this way we as women are Earth."

The Greening of Harlem is another solid example of ecofeminism founded by Bernadette Cozart in 1989. Bernadette Cozart was a professional gardener botanist and advocate that helped residents in New York regenerate and take responsibility for their communities. Run-down vacant lots were transformed into flower gardens and green spaces. School and community gardens became a reality through the Greening of Harlem Coalition. Majority of the people interested and actively participating in the project are women and the project proved economically beneficial and a great way for communities to contact nature and each other. He inspired a similar group in Detroit called the gardening angels to develop City Gardens as well as many other movements globally (MacGregor, 2017).

This great line of thinking has had an impact in highlighting the interrelationship between women and the environment and has made contributions beyond debate to involving women in public policymaking and environmental conservation. The contributions made by men and women alike to unpaid shared cooperation and local initiatives to solve environmental programs cannot be understated. Further research that is gender-sensitive and gender mainstreaming in all areas of life will prove beneficial to the design implementation and execution of environmental projects. As Vandana Shiva (Indian environmental activist) eloquently put it "We are either going to have a future where we might leave the way to make peace with the Earth or we are not going to have a human feature at all."

Bibliography

1. Australian National Center for Development Studies. Pacific Policy Paper 27. (1997). *The Ok Tedi Settlement: issues, outcomes, and implications*. edited by Banks G. and Ballard G. 294 pages
2. Agarwal, (2005). *Environmental Management* (pp. 1–79). New Delhi: APH Publishing.
3. Allan S.E., Smith B.W., Anderson K.A. 2012. *Impact of the Deepwater Horizon Oil Spill on Bioavailable Polycyclic Aromatic Hydrocarbons in Gulf of Mexico Coastal Waters*. Environmental Science & Technology 2012 46 (4), 2033-2039. DOI: 10.1021/es202942q
4. Alex, G., (2020). *Environmental Impact Assessment (EIA) Simplified @ GrG* (p. 104). New Delhi: BlueRose Publishers.
5. Arora-Jonsson, S., (2013). *Gender, Development, and Environmental Governance: Theorizing Connections* (pp. 1–73). London: Routledge.
6. Barrow, C.J. (2005). *Environmental Management and Development*. Routledge, London. 286 pages
7. Boamah, F. A., (2017). *Problems Associated with Material Management: Practice in the Construction Industry* (p. 61). Munich, Germany: GRIN Verlag.
8. Boylan, M., (2013). *Environmental Ethics* (p. 416). Florida: John Wiley & Sons.
9. Broughton E. 2005. *The Bhopal disaster and its aftermath: a review*. Environ Health volume 4, issue 6. https://doi.org/10.1186/1476-069X-4-6
10. Buckingham, S., (2005). *Gender and Environment* (pp. 1–84). London: Routledge.

11. Cassar, M., (2013). *Environmental Management: Guidelines for Museums and Galleries* (pp. 1–75). London: Routledge.
12. Chatterjee, S., (2004). *Applied Materials Management* (pp. 15–97). London: SAGE.
13. Chaytor, B., & Gray, K. R., (2013). *International Environmental Law and Policy in Africa* (pp. 31–83). Berlin: Springer Science & Business Media.
14. Dankelman, I., (2012). *Gender and Climate Change: An Introduction* (p. 312). London: Routledge.
15. Datta, A. K., (2008). *Materials Management: Procedures, Text and Cases* (pp. 3–70). New Delhi: PHI Learning Pvt. Ltd.
16. Dongo K., Tiembré I., Koné B.A., Zurbrügg C., Odermatt P., Tanner M., Zinsstag J. & Cissé G. 2012. *Exposure to toxic waste containing high concentrations of hydrogen sulphide illegally dumped in Abidjan, Côte d'Ivoire*. Environ Sci Pollut Res. Vol 19. 3192–3199 (2012). https://doi.org/10.1007/s11356-012-0823-2
17. Geert Van, C., & Denise, P., (2013). *Research Handbook on Environment, Health, and the WTO* (pp. 3–79). London: Edward Elgar Publishing.
18. Harris, F., (2005). *Global Environmental Issues* (pp. 1–19). Florida: John Wiley & Sons.
19. Jan, E., & Bert, B., (2012). *Activity-Based Cost and Environmental Management: A Different Approach to ISO 14000 Compliance* (pp. 23–60). Berlin: Springer Science & Business Media.
20. Jardins, J. R., (2012). *Environmental Ethics* (pp. 1–47). Boston: Cengage Learning.
21. Keller, D. R., (2010). *Environmental Ethics: The Big Questions* (pp. 25–83). Florida: John Wiley & Sons.
22. Kernohan, A., (2002). *Environmental Ethics: An Interactive Introduction* (p. 368). New York: Broadview Press.
23. King-Tak, (2009). *Environmental Ethics: Intercultural Perspectives* (pp. 11–115). Netherlands: Rodopi.
24. Kotze, L & Nel, J.G. 2009. *Environmental Management:* An introduction. Chapter 1 in Book: Environmental Management in South Africa. 2nd Edition. Editors: Strydom, H & King, N. Pages 1-33.
25. Lee, K. 1993. *Compass and Gyroscope: Integrity Science for Policy*

and the Environment. Island Press, Washington, D.C. 255 pages

26. Letcher, T. M., (2007). *Thermodynamics, Solubility, and Environmental Issues* (pp. 1–49). New York: Elsevier.
27. Levine A.G. 1982. *Love Canal: Science, Politics, and People. Lexington, MA*: D.C. Heath and Company. ISBN 978-0-669-05411-8. 263 pages
28. Ludivine, T., (2009). *Trade and Climate Change: A Report by the United Nations Environment Programme and the World Trade Organization* (pp. 1–76). World Trade Organization. London: UNEP/Earthprin.
29. MacGregor, S., (2017). *Routledge Handbook of Gender and Environment* (pp. 1–25). London: Taylor & Francis.
30. Medvedev Z.A. (1990). *The legacy of Chernobyl.* United States: Norton. 352 pages
31. Mishra, D. M., (2010). *Awareness Environmental Education IX* (pp. 1–93). New Delhi: S. Chand Publishing.
32. Mishra, R., (2009). *Materials Management* (pp. 4–43). New Delhi: Excel Books India.
33. Morris, A. S., (2004). *ISO 14000 Environmental Management Standards: Engineering and Financial Aspects* (p. p. 300). Florida: John Wiley & Sons.
34. Mukherjee, R., (2002). *Environmental Management and Awareness Issues* (pp. 1–60). New Delhi: Sterling Publishers Pvt. Ltd.
35. Nag, A., (2005). *Environmental Education and Solid Waste Management* (p. 106). New Delhi: New Age International.
36. Nath, B., (2009). *Environmental Education and Awareness-Volume II* (pp. 33–67). New Delhi: EOLSS Publications.
37. O'Riordan, T., (2014). *Environmental Science for Environmental Management* (pp. 1–63). London: Routledge.
38. OECD, (2007). *Policies for a Better Environment Progress in Eastern Europe, Caucasus, and Central Asia: Progress in Eastern Europe, Caucasus, and Central Asia.* Paris, France: OECD Publishing.
39. Petts, J., (2009). *Handbook of Environmental Impact Assessment: Volume 2: Impact and Limitation* (pp. 1–83). Florida: John Wiley & Sons.
40. Piasecki B. 1995. Corporate environmental strategy: *the avalanche of change since Bhopal*. New York (USA), John Wiley and Sons Inc. 163

pages

41. Piasecki, B., Fletcher, K. A., & Mendelson, F. J. (1999). Environmental management and business strategy: *leadership skills for the 21st century.* New York (USA) John Wiley & Sons Inc.
42. Rangarajan, M., (2009). *Environmental Issues in India: A Reader* (pp. 7–49). New Delhi: Pearson Education India.
43. Russo, M. V., (2008). *Environmental Management: Readings and Cases* (pp. 15–99). London: SAGE.
44. Sebastian-Coleman, L. (2018). Navigating the Labyrinth: *An executive guide to data management.* New Jersey. Technics Publications
45. Singh, D. A., (2008). *Materials Management* (pp. 4–93). New Delhi: Firewall Media.
46. Singh, V. P., (2005). *Toxic Metals and Environmental Issues* (p. 362). Delhi: Sarup & Sons.
47. Uberoi. (2004). *Environmental Management* (pp. 35–59). Delhi: Excel Books India.
48. Vijay, K., & Ramachandra, T. V., (2006). *Environmental Management* (pp. 1–32). New Delhi: The Energy and Resources Institute (TERI).
49. Wathern, P., (2013). *Environmental Impact Assessment: Theory and Practice* (p. 352). London: Routledge.
50. Welford, R., (2014). *Corporate Environmental Management 1: Systems and Strategies* (p. 280). London: Routledge.
51. Wiens T, Crist T.O., Day R.H., Murphy S.M., Hayward G.D. 1996. *Effects of the Exxon Valdez Oil Spill on Marine Bird Communities in Prince William Sound*, Alaska. Ecological Applications, Vol. 6, No. 3. pp. 828-841
52. Whitelaw, K., (2012). *ISO 14001 Environmental Systems Handbook* (pp. 1–22). London: Taylor & Francis.
53. Wood, C., (2014). *Environmental Impact Assessment: A Comparative Review* (pp. 1–64). London: Routledge.
54. World Bank. 2020. https://data.worldbank.org/indicator/SP.POP.TOTL

INDEX